Distributed Power Resources

Distributed Power Resources

Operation and Control of Connecting to the Grid

Ruisheng Li
XJ Electric CO., Ltd, Xuchang Henan, P.R.China

ACADEMIC PRESS
An imprint of Elsevier

Academic Press is an imprint of Elsevier
125 London Wall, London EC2Y 5AS, United Kingdom
525 B Street, Suite 1650, San Diego, CA 92101, United States
50 Hampshire Street, 5th Floor, Cambridge, MA 02139, United States
The Boulevard, Langford Lane, Kidlington, Oxford OX5 1GB, United Kingdom

Notices
Knowledge and best practice in this field are constantly changing. As new research and experience broaden our understanding, changes in research methods, professional practices, or medical treatment may become necessary.

Practitioners and researchers must always rely on their own experience and knowledge in evaluating and using any information, methods, compounds, or experiments described herein. In using such information or methods they should be mindful of their own safety and the safety of others, including parties for whom they have a professional responsibility.

To the fullest extent of the law, neither the Publisher nor the authors, contributors, or editors, assume any liability for any injury and/or damage to persons or property as a matter of products liability, negligence or otherwise, or from any use or operation of any methods, products, instructions, or ideas contained in the material herein.

Library of Congress Cataloging-in-Publication Data
A catalog record for this book is available from the Library of Congress

British Library Cataloguing-in-Publication Data
A catalogue record for this book is available from the British Library

ISBN 978-0-12-817447-0

For information on all Academic Press publications
visit our website at https://www.elsevier.com/books-and-journals

Publisher: Kattie Hammon
Acquisition Editor: Glyn Jones
Editorial Project Manager: Emma Hayes
Production Project Manager: Sruthi Satheesh
Cover Designer: Christian J. Bilbow

Typeset by SPi Global, India

Working together
to grow libraries in
developing countries

www.elsevier.com • www.bookaid.org

Contents

Chapter 1

Overview

For the clean, low-carbon, safe, and efficient new energy system, in order to make the most of wind energy, solar energy, and other renewable energy sources, connect more new energies to the grid, and improve the connecting capacity penetration of new energy, key technologies such as grid connection of new energy, energy storage control, electric vehicle connection, and microgrid control must be conquered. According to the mode of utilization, the grid connection of new energy can be divided into centralized utilization and distributed utilization. *Centralized utilization* refers to directly connecting a large-scale PV power station or a large-scale wind farm as a single large power source to the macrogrid, which is not discussed in this book. *Distributed utilization* refers to connecting the decentralized PV power generator and wind turbine to the distribution network by combining with the technologies of energy storage and electric vehicle connection, realizing local consumption of PV power and wind power (i.e., plug-and-play). This book mainly discusses the distributed utilization mode.

1.1 Concept and characteristics of distributed resources

1.1.1 Concept of distributed resources

The connected voltage class for centralized utilization of new energy is more than 35 kV, with a capacity over 10 MW, which facilitates the connection to the macrogrid; For distributed utilization, the distributed generating units are scattered around the consumers, which realizes local consumption. The surplus power is connected to the grid, with the connected power voltage lower than 10 kV and the capacity lower than 6 MW. Distributed generation has advantages of local consumption, low transmission and distribution loss, and low construction costs. PV power generator and wind turbine are connected to the grid by adopting the distributed utilization method, which is called *distributed generation*. IEEE 1547-2003, the IEEE Standard for Interconnecting Distributed Resources with Electric Power Systems, gives the following definition: distributed generation refers to a power generation facility that is connected to a local power system through the point of common coupling, which is a subset of distributed resources. As PV generating output varies along with the change of

Distributed Power Resources. https://doi.org/10.1016/B978-0-12-817447-0.00001-8

1

sun exposure time, and the wind power generating output varies along with the change of wind speed, the output of distributed generation varies along with the climatic changes such as sun exposure time and wind speed—that is to say, distribution generation features randomness, intermittence, and fluctuation, which are the greatest differences from traditional power generation.

AS the introduction of energy storage technology,the influence of the randomness, intermittence, and fluctuation of distributed generation have improved. The energy storage system can store excess energy in the grid during the low consumption period, and supply to the grid during the peak consumption period, which effectively improves the intermittence of distributed generation. IEEE 1547-2003 defines distributed resources as follows: Distributed resources are not directly connected to the macrogrid system, and the system consists of distributed generation and energy storage technology. The Advices on Provision of Distributed Resources Grid Connection Services, officially released by State Grid Corporation of China to the public in February 2013, clearly points out that the distributed power resources refer to power generation projects that are located close to the customers they serve, so as to the local consumption demands; in addition, the resources are connected to the grid with a voltage class not higher than 10 kV and less and with a total installed capacity of a single point not higher than 6 MW

The introduction of energy storage to the grid improves the randomness, intermittence, and fluctuation of distributed generation, which not only can help realize effective demand-side management, eliminate difference between peak and valley at day and night, and smooth the loads, but can also improve the availability of the power equipment effectively and reduce the power supply cost; besides, it can promote the application of renewable energy, improve the operating stability of the system, and offset load fluctuation.

Energy storage technologies applied in the field of grid system mainly include pumped storage, chemical batteries, supercapacitor storage, compressed air storage, and flywheel storage, while in the field of distributed generation, this mainly includes lithium battery, lead-acid battery, and supercapacitor storage. Besides, as the electric vehicle is getting popular, it not only can be used as a transportation media, but also serves as an effective controllable load: with vehicle-to-grid (V2G) technology, the energy generated by an electric vehicle can be transmitted to the grid, and the electric vehicle with such function can serve as an effective two-way controllable load of the grid. Electric vehicles are special carriers of these energy storage devices, and the charging devices in the charging station are equivalent to energy storage inverters, which realize energy exchange between the electric vehicles and the grids. The electric vehicle mode is relative to the travel time; therefore, uncertainty exists in terms of time and space dimensions.

Distributed PV generation, distributed wind power generation, energy storage, and electric vehicles are the trend for new energy utilization. They have a common feature: belonging to power electronic type resources (the biggest difference from traditional power sources). In addition, all of them have to be connected to the grid for field application, but differences exist in terms of time and space. This book mainly focuses on the power electronic type distributed resources, while other types of common distributed resources such as the synchronous generators and the induction machines are not discussed herein. This book, based on the friendly application of practical distributed resources grid connection, explains the controllable flexible grid connection technology for distributed PV generation, distributed wind power generation, energy storage, and electric vehicles, the adaptive grid active management technology, and the "plug-and-play" grid connection technology for distributed resources.

1.1.2   Characteristics of distributed resources

1.1.2.1   Power electronization

Distributed PV generation

Currently, grids are AC system. For connection of a PV generation system to an AC grid, DC power has to be converted into AC power. As shown in Fig. 1.1, the DC power generated by PV cells can be converted into AC power with a PV inverter, and then be connected to the AC grid. The PV inverter is a power electronic conversion device, which consists of an inverter bridge, a LC filter, and other elements, in which an inverter bridge is used for power conversion from DC power to AC power and an LC is used for filtering. The power flows unidirectionally.

In the future, both AC grids and DC grids will exist, constituting the AC&DC hybrid grids that can supply directly to both AC loads and DC loads. The AC&DC hybrid grids possess the advantages of both DC grids and AC grids. The DC load power will be supplied by the DC grid directly, skipping the AC-DC conversion procedure and reducing primary conversion loss. As shown in Fig. 1.2, DC power generated by a PV cell is connected to DC grids through a PV converter. (DC to AC inversion is required for connecting PV generated power to AC grids, and the device required for such AC grid connection is called an *inverter*, while DC to DC conversion is required for connecting PV generated power to DC grids, and the device required for such a conversion is called a *PV converter*.) The PV converter is a power electronic conversion device consisting of semiconductor for one-way DC/DC conversion. A PV

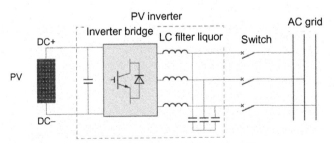

FIG. 1.1   PV generated power connected to AC grids through a PV inverter.

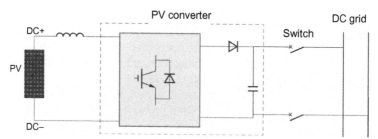

FIG. 1.2 PV generated power connected to DC grids through a PV converter.

converter generally adopts a boost circuit for DC to DC conversion, with power flowing unidirectionally from PV cells to DC grids.

Distributed wind power generation

Wind power generation is power generation that converts wind energy into electric energy. The wind generating set absorbs wind energy with a specially designed blade and converts wind energy to mechanical energy, which further drives the generator rotating and realizes conversion of wind energy to electric energy.

The commonly used wind power generation systems include the direct-driven wind power generating set and the double-fed wind power generating set; the direct-driven wind power generating set is connected to the grid through a full power converter, while the double-fed wind power generating set is connected to the grid through a double-fed converter.

Fig. 1.3 shows a direct-driven permanent magnet synchronous wind power generation system. For this system, wind energy drives the wind turbine rotating, which further drives the generator running, converting mechanical energy into electric energy. The stator of the permanent magnet synchronous generator outputs AC power with variable amplitude and frequency. By passing through an AC/DC rectifier, the AC power will be inverted into DC power, and then, with a DC/AC inverter, the output DC power will be inverted to AC power and connected to the AC grids. The power flows unidirectionally from the wind

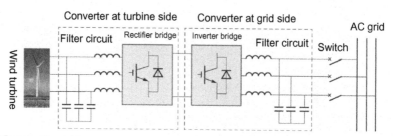

FIG. 1.3 Direct-driven permanent magnet synchronous wind power generation system.

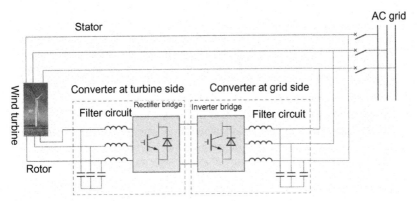

FIG. 1.4 Double-fed wind power generation system.

turbine to the AC grid. When it is only required to be connected to DC grids, the DC/AC inversion step can be omitted.

Fig. 1.4 shows the double-fed wind power generation system. Both the stator and the rotor of the double-fed generator can supply power to the grid, in which the rotator is connected to the grid through a converter, while the stator is connected to the grid directly. In case of speed change of the generator rotator, the converter will ensure the stator rotating magnetic field and the grid are in the same frequency by regulating the frequency of exciting current.

Energy storage

An energy storage system is equipped with the functions of energy storage, distributed resources output smoothing, and peak-load shifting, which can serve as both power source and load. As shown in Fig. 1.5, the energy storage battery is connected to the AC grid through the energy storage converter. If connecting to a DC grid is required, a DC/DC converter should be used for voltage amplitude transform, so as to realize the bidirectional flow of energy between the energy storage systems and the grid.

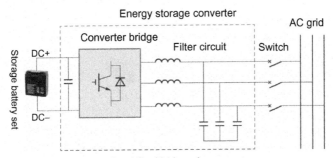

FIG. 1.5 Stored energy connected to AC grids through a converter.

Electric vehicle

In the future, the electric vehicle can be involved in balancing the local energy of the grid by serving as a random load or a distributed energy storage, realizing two-way interaction with the grid, and supporting grid-connection and power consumption of the large-scale distributed resources. Fig. 1.6 shows the connection of electric vehicle to the AC grid through a converter, with power flowing bidirectionally between the battery and the grid.

The PV generation, wind power generation, energy storage devices, and electric vehicle charging/discharging are connected to the grid via the power electronic converter. All the inverter, the converter and the charger can realize power conversion, which is used to meet the power conversion demands in different application contexts. All of them adopt DC/DC and DC/AC power conversion devices, as shown in Fig. 1.7.

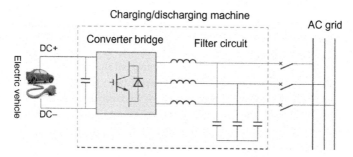

FIG. 1.6 Grid connection of electric vehicle (V2G).

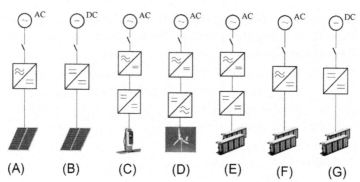

FIG. 1.7 Different types of power electronic resources. (A) PV inversion (DC/AC); (B) PV conversion (DC/DC); (C) Charger (DC/DC++DC/AC); (D) Wind turbine conversion (AC/DC+DC/AC); (E) Energy storage conversion (DC/DC++DC/AC); (F) Energy storage conversion (DC/AC); (G) Energy storage conversion (DC/DC).

1.1.2.2   PCC voltage change

When the distributed resources are connected to the grids, the flow distribution of the traditional grids will change, which will result in changes to the steady voltage at PCC. PCC voltage variation is related to the PCC position and capacity of the distributed resources. Active power input may result in PCC voltage boosting. The response time for voltage anomaly and the time for grid disconnection of distributed resources are stipulated in applicable standards, which are as shown in Table 1.1. As we can see from Table 1.1, for grid connection of distributed resources, the requirements for abnormal voltage response are strict.

When $U > 1.1U_n$, the distributed resources will be disconnected from the grid within the stipulated time, in which requirements set out in Standard [1] and [2] in Table 1.1 are extremely strict. When $1.1U_n < U < 1.2U_n$, the grid disconnection time will be a 60(100) grid cycle; when $U \geq 1.2U_n$, the grid disconnection time will be a 10 grid cycle. In this way, the adverse impact imposed on the grid by the grid connection of distributed resources can be avoided.

Like the abnormal voltage response, for the purpose of ensuring safe grid connection and avoiding islanding, both national standards and international standards stipulate the requirements for abnormal frequency response after grid connection of distributed resources. For example, in case of the upper limit exceeding 0.5 Hz or the lower limit exceeding 0.7 Hz, the distributed resources will be disconnected from the grid within the stipulated time (i.e., 6 grid cycles). Table 1.2 shows the nominal voltage and frequency of the grid under normal operation of the distributed resources.

1.1.2.3   Randomness

Requirements for grid connection of different distributed resources are the same. But differences in terms of time, space, and power flow exist among different types of resources. In terms of time, PV generating outputs vary along with alternation of day and night as well as the sunset time; wind power generating outputs vary along with the climate change and wind speed; energy storage outputs varies as the peak-load shifting curve changes; and electric vehicle outputs varies along with the traveling behavior. In terms of space, positions of PV generation, wind power generation, and energy storage are fixed, whereas the position of the electric vehicle changes as the vehicle travels. In terms of power flow, the PV generation and the wind power generation supply power to the grids, with power flowing unidirectionally, while the energy storage and the electric vehicle can supply power to the grid and also can receive power supplied by the grid, with power flow bidirectionally.

1.1.2.4   Consumption difficulty

Electrical characteristics for grid connection of different distributed resources are the same, but their requirements for grid connection are different. As the distributed PV generation and the wind power generation pursue high energy

TABLE 1.1 Requirements for overvoltage response in different applicable standards

Overvoltage range (%U_n)	Maximum disconnecting cycle count N^a	Applicable standard
110 < U < 120	60 [1],100 [2] ≥500 [8]	[1] IEEE 1547–200, National Standard for Interconnecting Distributed Resources with Electric Power Systems
U ≥ 120	10 [1],10 [2],6 [13]	[2] Q/GDW 564–2010, Technical Guideline for Electric Energy Storage System Interconnecting to Distribution Network
110 < U < 135	100 [3–7,9,10] 120 [11]	[3] Q/GDW 480–2010 Technical Rule for Distributed Resources Connected to Power Grid
U ≥ 135	10 [3, 9]; 2.5 [4–7, 10]; 3 [11]	[4] NB/T 32004–2013, Technical Specification of Grid-Connected PV Inverter
		[5] GB/T 20046–2006, Photovoltaic (PV) Systems—Characteristics of the Utility Interface
		[6] GB/T 19939–2005, Technical Requirements for Grid Connection of PV System
		[7] Q/GDW 617–2011, Technical Requirements for Connecting Photovoltaic Power Station to Power Grids
		[8] GB/Z 19964–2005, Technical Requirements for Connecting Photovoltaic Power Station to Power System
		[9] GB/T 29319–2012, Technical Requirements for Connecting Photovoltaic Power System to Distribution Network
		[10] CGC/GF001: 2009, Technical Specification and Test Method of Grid-Connected PV Inverter Below 400 V
		[11] IEC 61727–2004, Photovoltaic (PV) Systems—Characteristics of the Utility Interface
120 ≤ U ≤ 130	≥2.5 [8]	[12] V-E-R-N4105: 2011-08, Power Generation Systems Connected to the Low-Voltage Distribution Network
110 < U < 115	6 [12]	[13] BDE, Generating Plants Connected to the Medium-Voltage Network
U > 115	6 [12]	

^{a}N stands for the grid cycle count; the values are calculated based on 60Hz stipulated in the international standard and on 50Hz stipulated in the national standard.

TABLE 1.2 Nominal grid voltage and frequency under normal operation of the distributed resources

Voltage range (%U_n)	Frequency	Applicable standard
$85 < U < 110$	f_{nom}	[1] IEEE 1547.2003, *IEEE Standard for Interconnecting Distributed Resources*
U_{nom}	$f_{nom} - 0.7\,Hz < f$ $f < f_{nom} + 0.5\,Hz$	[2] NBT 32004–2013, *Technical Specification of Grid-Connected PV Inverter* [3] GB-T 20046–2006, *Photovoltaic (PV) Systems—Characteristics of the Utility Interface* [4] GB-T 10030–2005, *Technical Requirements for Grid Connection of PV System* [5] GB-Z 19964–2005, *Technical Requirements for Connecting Photovoltaic Power Station to Power System* [6] GB/T 29319–2012, *Technical Requirements for Connecting Photovoltaic Power System to Distribution Network* [7] IEC 61727–2004, *Photovoltaic (PV) Systems—Characteristics of the Utility Interface* [8] VDC-R-N 4105: 20-1-08 *Power Generation Systems Connected to the Low-Voltage Distribution Network* [9] Medium-Voltage Grid Connection Standard of Germany, BDEW, *Generating Plants Connected to the Medium-Voltage Network*

U_{n-m}, nominal voltage amplitude of the grid; f_{n-m}, nominal frequency of the grid.

yield, therefore, higher capacity penetration is preferred. However, if the large-scale grid connection and the load variability are not properly restrained, severe voltage deviation, voltage fluctuation, or even out-of-limit voltage may occur, which may adversely affect the grid safety. This results in the contradiction between high capacity penetration and grid safety. If the capacity penetration is increased for high power yield, the over-high capacity penetration may result in inverter (converter) trip due to overvoltage, which leads to frequent connection and disconnection of the distributed power resources and further reduces the power yield by the distributed resource. The energy storage unit may effectively offset the intermittence of the distributed generation, but the spatial difference, the unrestrained charging and discharging, as well as the load variability of electric vehicles can also result in voltage deviation and voltage fluctuation, thus influencing the grid safety.

1.2 Requirements for grid connection of distributed resources

1.2.1 Safe grid connection

When the distributed resources are connected to the grid,connection safety , especially an islanding event during power supply to the loads due to power supply failure will be considered. It is very important to ensure a safe grid connection of the distributed resources and ensure the personal safety of the maintenance personnel. Currently there are two methods for islanding detection. One is that the detection unit is arranged within the inverter, which can be classified into the passive method and the active method. For the passive method, islanding events are identified based on failures of voltage, frequency, phase, and harmonics, but a dead zone exists, while for the active method, islanding events are identified by imposing disturbance to the amplitude, the frequency and the initial phase of the output current. In this method, power disturbance, frequency disturbance, and phase deviation are widely applied. When multiple inverters are running in parallel, the detection results of the active method may be inaccurate due to out-sync disturbance. The other method is that the detection unit is arranged outside of the inverter, and islanding events are identified based on the status of the breaker detected by communication means, or a special anti-islanding device should be configured. The working principle for islanding detection of this method is to break the power balance by controlling the switch or imposing disturbance to the loads. This method also has disadvantages: it is complicated and relies on the communication system for breaker status detection, and it requires high automation level of the grid; in addition, the special anti-islanding device is operated manually. Therefore, an active measure against islanding without relying on the inverter and the communication system is required to ensure safe grid connection of the distributed resources.

1.2.2 Information interconnection

Associated interconnecting devices are necessary for grid connection of the distributed resources. Currently, there is plenty of information about the interconnecting device, and the applicable regulations are diversified. Information interconnection is required, so as to realize information exchange and "plug-and-play." To meet the requirements of the operation and control mode of the distributed resources, the IEC 61850-90-7 Distributed Resources System has built the model in accordance with IEC61850. It supports emergency control of the distributed resources, voltage and reactive power control, frequency control, voltage management, and other functions, realizing information exchange of the distributed resources. Characteristics of the information interconnection technology are "distributed resources + Internet.". Commissioning and calibration of the distributed resources is completed with mobile terminals, which realizes noncontact commissioning, detection, and diagnosis, and ensures personal and equipment safety.

1.2.3   Control of high capacity penetration

When the capacity penetration of the distributed resource reaches up to 15%–20%, power balance, safety, and stability of the grid will be affected. In order to improve the capacity penetration of the distributed resources, the control system will carry out active management over the grid with large-scale connected distributed resources by adopting the communication and power electronic technologies; the distributed resources will meet the requirements of the coordination control system and will respond quickly to the scheduling signal sent out by the control system.

1.2.4   Unification of power electronic resources conversion

The distributed resources belong to power electronic resources, which can be classified as inverters, converters, and chargers based on their application contexts. In engineering application, the power of electronic resources is specially designed according to the actual application conditions, which results in the following disadvantages: different topological structures, different electrical interfaces, poor replaceability in case of equipment damage, severe wear to switches of traditional two-level DC/AC and DC/DC, as well as low efficiency and low power density of the complete generator. Currently, it is hard to solve the technical difficulties for AC-to-grid connection, such as PCC voltage fluctuation and frequency deviation, grid fault ride-through, load imbalance reduction, seamless connection and disconnection of the microgrid, and auto interconnection of multiunits without communication lines. For the AC&DC hybrid microgrid system that will be applied in the future, technical difficulties also exist, such as PCC power fluctuation and voltage oscillation on the DC grid, voltage supports on DC grid under the islanding operation mode, and DC auto interconnection without a communication line. Therefore, it is necessary to unify the conversion of different types of the distributed resources. The three-level DC/DC and DC/AC with a high power density, unified interface, able to carry out auto parallel connection, will be used to structure the resources conversion device, so as to realize "plug-and-play" of the equipment connected to the distributed resources.

1.3   Key technologies of grid connection of distributed resources

1.3.1   Low-frequency power injection technology for active islanding detection

For grid connection of distributed resources, in order to ensure grid safety, the connected distributed resources will be equipped with islanding protection. In case of an islanding event, the islanding protection will be activated, disconnecting the distributed resources from the grid, so as to ensure personal and equipment safety. Currently, the following problems exist in grid connection of distributed resources: (1) parallel connection of multiple distributed resources may fail the islanding protection; (2) when DG adopts the characteristics of a virtual synchronous generator, the islanding protection lapses; and (3) islanding protection are not provided within the two-way charger and thus a low-frequency power injection type active islanding detection technology without depending on the distributed resources and on the communication system is required. A safety interconnection device integrated with the functions of switching, protection, measurement, control, and communication will be provided so as to ensure the safety of electrical interconnection and facilitate grid connection, realizing "plug-and-play" of the grid-connected distributed resources.

1.3.2 Auto overvoltage and power control technology (U/P)

The unrestrained large-scale connection of distributed resources to the grid and loads variability may result in severe voltage deviation, fluctuation, and even overvoltage, which may further result in inverter disconnection. When the distributed resources are working under PQ mode, the voltage/power control technology will be adopted. With this technology, the distributed resource may adjust its output according to the PCC voltage, which solves the problem of power generation failure due to distributed voltage disconnection from the grid caused by voltage boosting due to overlarge active power output of distributed resources; after the voltage resumes its normal value, the distributed resources will recover to normal operation, which optimizes the time for grid connection of distributed resources, realizing operation with the maximum generating capacity penetration.

1.3.3 Auto overfrequency and power control technology (f/P)

In a general condition, after grid connection of the distributed resources, the resources will run under the maximum power point tracking mode, without participating in dynamic frequency and voltage regulation of the grid. With the auto overfrequency and power control technology, the distributed resources will actively participate in frequency and voltage control of the grid, which improves the friendliness of the grid-connected distributed resources. During grid-disconnected operation of microgrids, the off-grid energy of microgrids will be kept in balance in a real-time manner. When the battery is fully charged (i.e., under the high state of charge), excess energy can neither be stored nor consumed by loads; in this case, the excess energy may break the energy balance and result in microgrid failure. With the auto over-frequency and power control technology, the output of the distributed resources can be properly controlled, so as to ensure energy balance in the grid-disconnected operation mode.

1.3.4   Presynchronous grid connection technology

When the distributed resource is operating in the grid-disconnected state and works as a voltage source such as the light storage machine and the microgrid system, voltage of the distributed resources generally deviates from the voltage at grid side (phase, amplitude, and frequency). When distributed resources are connected to the grid directly without being subject to synchronous control, the small voltage difference and small phase difference will impose on small connecting impedance, which will generate a large surge current and even lead to grid connection failure and equipment damage. By adopting the presynchronous grid connection technology, the amplitude and phase differences will decrease gradually, which ensures that the voltage amplitude and phase of the voltage sources are consistent with those at grid side, and realize auto grid connection with "zero surge."

1.3.5   Self-optimization virtual synchronous generator technology

The distributed resources adopt power electronic type resources conversion, and the power electronic resources are controlled by the digital circuit, with fast transient response speed and lower inertia, which cannot participate in frequency and voltage regulation of the grid. In the case of low DG capacity penetration, the grid will supply stable voltage and frequency; in the case of high DG capacity penetration, too many inertia-free distributed resources may affect the stable operation of the grid. In this case, DG with inertia can be used for grid regulation, so as to realize a DG "plug-and-play" grid connection with high capacity penetration. For example, when connecting DG to a grid with high capacity penetration where the system frequency deviation is large, DG with rotational inertia will cause rotational inertia increase of the whole grid system and smooth system frequency change, while damped DG may increase the damping of the whole grid system, so as to shorten the transient process for system frequency change. In microgrid applications, DG with inertia will enhance the effect of rotational inertia of the whole microgrid and significantly improve the operation stability of the microgrid during grid-disconnected operation; in addition, it may tackle the technical difficulties such as seamless on-off grid switching and overvoltage during the planned islanding and unplanned islanding of microgrid. With the virtual synchronous generator technology, friendly grid connection of distributed resources, "plug-and-play" microgrid connection, and grid connection of the distributed resources with high capacity penetration can be realized.

   The rotational inertia and damping of the traditional synchronous generator are fixed. After design and manufacture of a synchronous generator, the rotational inertia and damping of such a generator cannot be changed. With self-optimization virtual synchronous generator technology, the characteristics of the power electronic resources of being flexible and controllable can be particularly advantageous. The generator possesses the rotational inertia and the damping that the synchronous generator possessed, participating in grid regulation, and overcomes the shortcoming of the synchronous generator with fixed rotational inertia and damping. In the case of a large disturbance to the voltage frequency, the rotational inertia and damping will increase, while in case of small disturbance to the voltage frequency, the rotational inertia and the damping will decrease. The rotational inertia and the damping can change along with the magnitude of disturbance, realizing self-adaption to the frequency change.

1.3.6 AC&DC hybrid microgrid coordination control technology

The AC&DC hybrid microgrid has both AC bus and DC bus, which can supply directly to AC loads and to DC loads. It can exert the advantages of the AC microgrid and the DC microgrid; therefore the AC&DC hybrid microgrid coordination control technology is absolutely necessary. During grid-connected operation of the AC&DC hybrid microgrid, the power flow between the AC bus and the DC bus should be controlled; during grid disconnected operation, based on the AC bus or the DC bus, the voltage of relevant buses should be controlled, so as to ensure stable grid-disconnected operation of the AC&DC hybrid microgrid.

1.3.7 Communication line-free interconnected microgrid control technology

The current microgrid has a complex structure configured with many control devices. The distributed resources, energy storage devices, and loads are under the centralized management of the Microgrid Control Center, realizing off-grid energy balance of the microgrid. Considering the construction costs and economic efficiency, this kind of microgrid is not appropriate for commercial applications. In this regard, with the communication line-free interconnected microgrid control technology, a "plug-and-play" microgrid adopting the simplest possible physical structure without depending on communication and control devices, and which can realize self-parallel connection only with the energy storage device and DG, is required, so as to reduce the grid construction costs to a level similar to that for distributed resources construction and facilitate microgrid construction, popularization, and application.

1.3.8 Active grid protection control technology

After grid connection of the distributed resources, the power flow on the grid changes from one-way flow to two-way flow, which will influence the processes of grid protection, distribution automation, and failure treatment. Main influences are as follows: sensitivity decrease of the terminal fault current protection, false action of adjacent line protection, reconnection failure, and so on.

After grid connection of the distributed resources, a new relay protection technology will be introduced to the nonradioactive (networked) grid of the distributed resources according to the layout of the cables and overhead lines of the grid, so as to adapt to the active grid in the future. The relay protection technology of the transmission grid is introduced to the grid, and current differential protection and directional comparison protection are adopted for relay protection of the active grid, which realizes fast and accurate failure locating and failure isolation on the grid.

1.4 Application mode of distributed resources

1.4.1 Grid connection mode of distributed resources

For grid connection of the distributed resources, there are two connecting voltage classes (i.e., 10 kV and 380 V/220 V), and selection of these two connecting modes is subject to the connecting capacity. As shown in Table 1.3, when the connecting capacity is higher than 400 kW, the three-phase connecting mode with connecting voltage class of 10 kV should be selected; when the connecting capacity falls between 8 and 400 kW, the three-phase connecting mode with connecting voltage class of 380 V should be selected; when the connecting capacity is lower than 8 kW, the single-phase connecting mode with a connecting voltage class of 220 V should be selected. According to the actual connecting conditions, the connecting capacity can be divided into dedicated line connection and T-type connection. Fig. 1.8 shows the typical connection mode for the distributed resources. For T-type connected capacity, the 10 kV voltage classes will meet the transformer capacity requirement, and the connecting capacity will not be higher than 3 MW; the 380 V voltage class will meet the transformer capacity requirement, and the connecting capacity will not be higher than 50 kW.

TABLE 1.3 Typical grid connection mode and capacity of distributed resources

Voltage		<8kW	8–50kW	30–400kW	400–3000kW	2000–6000kW
				Capacity		
10kV		–	–	–	T-type connection	Dedicated line
380/ 220	380V	–	T-type connection	Dedicated line	–	–
	220V	T-type connection	–	–	–	–

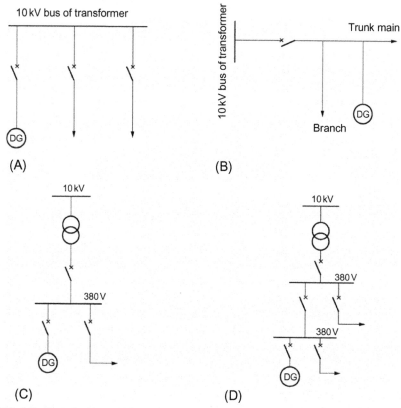

FIG. 1.8 Typical grid connection patten of distributed resources. (A) 10 kV dedicated line connection; (B) 10 kV T-type connection; (C) 380 V dedicated line connection; (D) 380 V T-type connection.

1.4.2 Mode of PC and energy storage AIO machine

In order to increase the connecting capacity penetration of the distributed resources, there are mainly two modes for grid connection of the distributed resources: one is regional distributed generation connection, which generally has large capacity (higher than 50 kW); the other is PV connection for household use, which has small capacity (lower than 20 kW) and meets the consumption requirements of one or several households. Through the PV and energy storage AIO machine, PV generation for household use can meet the basic power demands of home lighting and daily electric appliances. There are three structures of the PV and energy storage AIO machine: shared AC bus, shared PV bus, and shared DC bus. For the shared DC bus structure, as shown in Fig. 1.9C, PV cell is connected to the DC bus through a one-way DC/DC while energy storage device is connected to the DC bus through a two-way DC/DC,

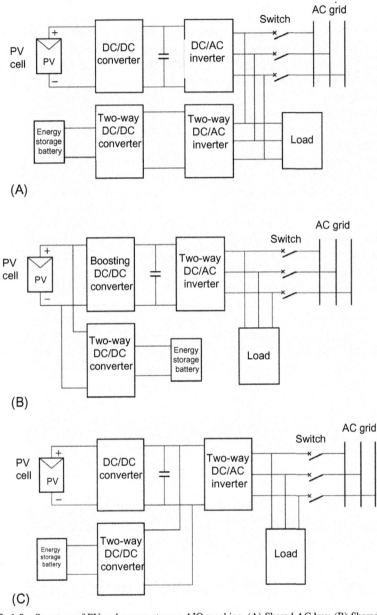

FIG. 1.9 Structure of PV and energy storage AIO machine. (A) Shared AC bus; (B) Shared PV bus; (C) Shared DC bus.

and the DC bus is connected to the grid through a two-way DC/AC. This structure has advantages of less charging stage, high efficiency, and low cost, which is a commonly applied structure AIO.

1.4.3 Mode of microgrid

Another good power utilization mode in the distributed generation area is the microgrid. The microgrid is an autonomous system with functions of self-control, management, and protection, consisting of distributed generators, energy storage devices, loads, and control devices. It has two operation modes: grid-connected mode and grid-disconnected mode. The microgrid technology is a technical approach to effective utilization of distributed generation. Micro-grids can be divided into AC microgrids, DC microgrids, and AC&DC hybrid microgrids. Fig. 1.10 shows the structure of AC&DC hybrid microgrid. In the AC&DC hybrid microgrid, there are two points of common coupling (PCC), in which PCC1 is an AC PCC, while PCC2 is a DC PCC. In the figure, the AC microgrid is connected through PCC1. In the AC microgrid, the distributed gen-erator is connected to the AC bus via a one-way DC/AC, while the energy stor-age is connected to the AC bus via a two-way DC/AC. The AC microgrid has two operation mods (i.e., the grid-connected mode and grid-disconnected mode); the grid connected through PCC2 is the DC microgrid. In the DC micro-grid, the distributed generator is connected to the DC bus via a one-way DC/DC, while the energy storage is connected to the AC bus via a two-way DC/DC. The AC microgrid has two operation modes: the grid-connected mode and the grid-disconnected mode. In the AC&DC hybrid microgrid, bidirectional power flow between the AC bus of the AC microgrid and the DC bus of the DC microgrid is realized through a DC/AC coordinate controller, and the AC&DC hybrid micro-grid also has two operation modes (i.e., the grid-connected mode and the grid-disconnected mode).

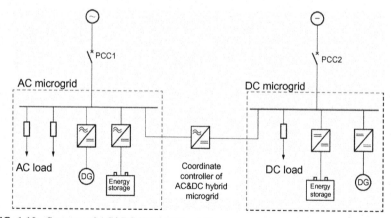

FIG. 1.10 Structure of AC&DC Hybrid Microgrid.

Chapter 2

Grid-connected power conversion of distributed resources

All distributed resources have a common feature: power electronic resources. For grid connection of distributed resources, DC/AC or DC/DC conversion is realized through power conversion. No matter if the grid connection is realized with a voltage source or with a current source, the basic DC/AC and DC/DC converter is used for power conversion. As the traditional DC/AC and DC/DC converters adopt the two-level technology, LC filtering will be designed for the outputs; for multilevel technology with 10 levels and above, filtering is not required for the power frequency AC output. Multilevel technology with four levels and above is not competitive in terms of costs, but multilevel technology with three levels can be applied. Advantages include that the output is close to the sine wave, with low harmonic content and high conversion efficiency, and its influence on the system is minor. The filter inductance of three-level technology is small, which can reduce costs; aside from this, the device with this technology is featured by small structure, lightweight, and high power density of the complete generator. Advantages of power electronic resources conversion with three-level technology are obvious, and ideal AC output can be obtained with simple filtering. This chapter takes DC/AC and DC/DC conversion with three-level technology as an example, providing a brief introduction to power electronic resources conversion.

2.1 Three-level DC/AC resources conversion

Currently, the DC/AC converter adopts the power conversion technology of pulse width modulation, developing from two-level to multilevel. Level number refers to the voltage class of the DC/AC output (before filter inductance) relative to the phase voltage of the neutral point of the DC bus. Voltage class no less than 3 belongs to multilevel. Voltage class of $\pm U/2$ belongs to two-level, and the line voltage will be 0, $\pm U$. If the phase voltage is 0 and $\pm U/2$, it belongs to three-level, and the line voltage will be 0, $\pm U/2$, and $\pm U$. When the phase

Distributed Power Resources. https://doi.org/10.1016/B978-0-12-817447-0.00002-X

19

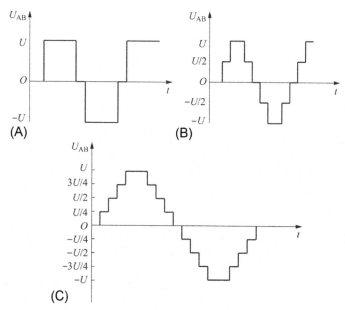

FIG. 2.1 Voltage diagram of two-level, three-level, and five-level circuits. (A) Two-level; (B) three-level; and (C) five-level.

voltage is 0, $\pm U/4$, and $\pm U/2$, it belongs to five-level, and the line voltage will be 0, $\pm U/4$, $\pm U/2$, $\pm 3U/4$, and $\pm U$, as shown in Fig. 2.1.

Compared with two-level, the three-level has the following advantages: (1) low harmonic content, low filter inductance; (2) low switching frequency, low switching loss, small voltage change ratio (dU/dt) during switching operation, and small electromagnetic interference; (3) high system efficiency and easy modular parallel connection. The higher the level number is, the closer the output voltage waveform is to the sine wave, but the topological structure and the control system are complicated. Therefore, a three-level DC/AC power module that realizes auto connection of the power module for large capacity resources conversion and constitutes full power coverage DC to AC power electronic resources conversion is the reasonable and economical technical route.

2.1.1 Topological structure of three-level DC/AC

Currently, there are mainly three types of three-level topological structures.

(1) The neutral point clamped (NPC) three-level topological structure is also called a *diode-clamped three-level topological structure*, which was first proposed by a Germany scholar Holtz in 1977. As shown in Fig. 2.2A, two capacitors in series connection are arranged at the DC side, and two clamping diodes and four switch tubes are arranged between the bridge

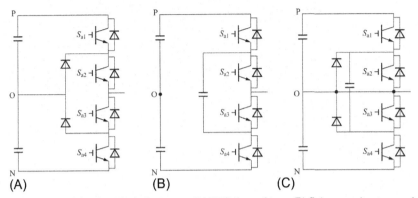

FIG. 2.2 Three-level topological structure. (A) NPC clamped type; (B) flying capacitor type; and (C) hybrid clamped type.

arms. The voltage at DC side is U_{dc}, voltage imposing on each capacitor is $U_{dc}/2$, and output phase voltage is $U_{dc}/2$, 0, and $-U_{dc}/2$, respectively. Under the effect of the clamped diode, voltage imposing on each switching element is restrained to $U_{dc}/2$; besides, under the blocking state, each clamped diode bears the same voltage. This structure is a typical NPC structure, which belongs to I-type topology.

(2) The flying capacitor three-level topological structure is as shown in Fig. 2.2B, which is based on the diode-clamped three-level circuit and substitutes the clamped diode with the capacitor, realizing voltage clamping for the switching tube.

(3) The hybrid clamped three-level structure is as shown in Fig. 2.2C, which combines the NPC topological structure and the flying capacitor topological structure.

In addition to the typical NPC three-level topology, there are derived topological structures, including an I-type voltage divider with a capacitor, an I-type active NPC, and T-type topological structures (classified into two types: common collector and common-emitter), as shown in Fig. 2.3.

2.1.2   Comparison of different NPC topologies of three-level DC/AC

As shown in Fig. 2.3A, the NPC of an I-type voltage divider with a capacitor can avoid direct connection to the center of DC side, so as to reduce common mode leakage current. As shown in Fig. 2.3B, an I-type active NPC adopts the active clamped method, substituting an active switch tube with an antiparallel diode for the clamped diode of the traditional typical NPC structure. With this structure, the control is more flexible, since a continuous current circuit is added, and the control flexibility of the system is improved, but the control method is complicated. Fig. 2.3C shows the T-type topological structure. In

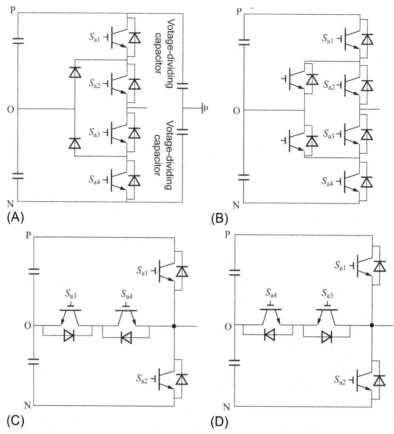

FIG. 2.3 NPC three-level derived topological structure. (A) I-type voltage divider with capacitor; (B) I-type active NPC; (C) T-type common collector; and (D) T-type common emitter.

the T-type topological structure, the new power switching tube RB-IGBT with reverse blocking function is used to substitute for the switching tube in reversed series connection, which can facilitate loss, improve inversion efficiency, and reduce the volume of the resources conversion device. The topological structure is as shown in Fig. 2.4.

Main differences between T-type and I-type are as follows:

(1) *Voltage resistance.* In principle, the voltage resistance of I-type circuit is better than that of T-type circuit, but from the view of practical application, voltage resistance of these two types are similar.

(2) *Loss.* When switching frequency is less than 20 kHz, loss of T-type is less than that of I-type.

(3) *Element quantity.* Compared with I-type, T-type has fewer elements (two diodes).

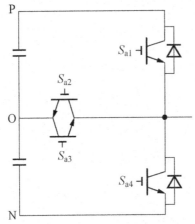

FIG. 2.4 T-type topological structure of RB-IGBT with the reverse blocking function.

(4) *Control sequences.* For I-type, the external tube should be disconnected first, then the internal tube, so as to prevent imposing of bus voltage on the internal tube and resulting in damage. For T-type, there is no sequence requirement.

(5) *Conversion path.* The conversion path of I-type topology can be classified into short conversion path and long conversion path. Where discrete models are used, the problems of stray inductance and voltage spike must be considered; when the conversion paths of the internal tube and the external tube of T-type topology are the same, there is only one conversion loop. For details regarding the conversion path, please refer to Fig. 2.5.

2.1.3 Comparison between flying capacitor type and typical NPC

Compared with the typical NPC three-level topological structure, the flying capacitor topology has the following advantages: (1) rectify the shortage of too many diodes; (2) there are multiple combinations of switching mode, which can balance the voltage on the capacitor; (3) with the flying capacitor, the harmonic distortion rate of the output voltage and the dU/dt of the switching elements are relatively small. In addition, when the switching element is under the blocking condition, better voltage balance can be realized. The disadvantage of the flying capacitor topology is as follows: (1) a large amount of clamped electrolytic capacitors are required, and the system is in large size, which impedes system integration; (2) frequent charging and discharge will result in short service life of the clamped capacitor, with poor reliability.

Comparison results between the NPC three-level topology and the flying capacitor topology are as shown in Table 2.1.

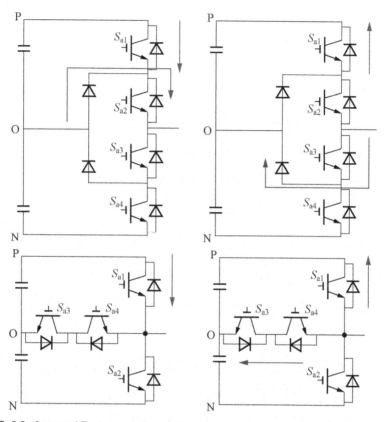

FIG. 2.5 I-type and T-type conversion path.

TABLE 2.1 Comparison between NPC three-level topology and flying capacitor topology

Topological structure	NPC	Flying capacitor	Topological structure	NPC	Flying capacitor
Number of switching tube in each phase	4	4	Loss distribution	Uneven	Even
Number of clamped-diode in each phase	2	0	Reliability	High	Low
Number of capacitance in each phase	0	1			

2.1.4 Comparison of hybrid clamped type and typical NPC

For the typical NPC topology, the hybrid clamped type adds a clamped capacitor, so that the second and the third power tubes of each bridge arm cannot be connected simultaneously. Otherwise, the clamped capacitor will be shortcut. In this case, there will be two statuses for obtaining zero level—that is, the switching status of each bridge arm will increase from three to four. In addition, the voltage space vector will increase to 64 types, which improves the flexibility of the system control while increasing the complexity of the system.

2.1.5 T-type circuit improvement

After taking all factors into consideration, T-type structure is preferred. However, the following problems exist in the traditional T-type circuit: (1) The common mode filtering circuit is not decoupled with the differential mode filtering circuit, and the resonant frequency of the common mode filter is high, which is easily subject to resonance. (2) When adopting the space vector pulse width modulation (SVP-WM) algorithm, large zero-sequence current will pass through the neutral line, resulting in system loss increase.

A structural scheme based on decoupling of common-mode and differential-mode filters is adopted: the LCL filter capacitor is divided into two parts in parallel connection, the common point of the filter capacitor is connected with small capacitance to the neutral point of the capacitor at DC side, so as to reduce the coupling degree between the common-mode filtering circuit and the differential-mode filtering circuit, and facilitate separate design, as shown in Fig. 2.6. This topological structure will not affect the filtering effect of LCL filter; the common-mode filtering circuit can be designed with small capacitance, so as to increase the loop impedance, reduce neutral line current, reduce

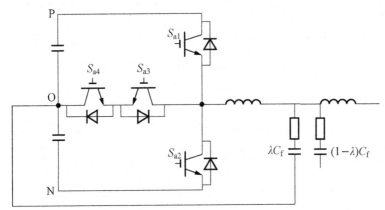

FIG. 2.6 Topology diagram of decoupling structure of common-mode and differential-mode filters.

system loss, improve the resonant frequency of the common-mode circuit, and avoid frequency section easy subject to resonance.

2.1.6 Neutral-point potential balance control

The inherent problems of unbalance potential at neutral point exist in the three-level DC/AC topology; this problem can be classified into two categories: one is potential fluctuation at the neutral point, which will result in low-order harmonics in the output voltage, and reduce output power quality; the other is potential deviation at neutral point, which will result in output voltage waveform distortion. Sometimes the over-high capacitor voltage may even damage the capacitor and the switching element at DC side.

To restrain the potential imbalance at the neutral point, first of all, the problem of potential deviation at the neutral point should be settled, and then control the potential fluctuation at the neutral point. There are two modes of potential balance control at the neutral point: hardware control and software control.

2.1.6.1 Hardware control

Fig. 2.7 shows neutral potential balance circuit. Compared with a common circuit, S_{a3} is added. This circuit can effectively restrain the potential deviation at the neutral point. The hardware control is more accurate with easier procedures, but a large number of elements should be added to the circuit, which increases the complexity and the costs of the circuit.

2.1.6.2 Software control

Carry out neutral point potential balance control with software, such as the method of balance factor. Based on the SVPWM algorithm, change the acting time allocation of the small redundant vector, so as to realize neutral point potential balance control. The SPWM algorithm is adopted, such as the specific harmonic elimination modulation based on hysteresis control, and the PWM method based on zero sequence component injection. The zero-sequence

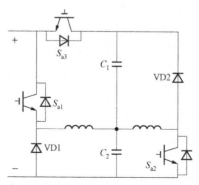

FIG. 2.7 Main neutral-point potential balance circuit.

component injection method, from the aspect of modulating wave, calculates the zero-sequence component for reaching neutral-point potential balance in a carrier wave cycle based on the influence imposed on the neutral point potential by the zero-sequence component, and realizes neutral point balance through the zero-sequence component injection method.

This method does not need vector partition and calculation of vector acting time, which facilitates engineering realization. For the T-type three-level inverter control method based on model predictive control, a specific performance indicator function should be established, and then the optimal switching status should be selected and imposed on the inverter. This method omits the traditional modulation steps and only uses a predicting controller, which is featured by flexible control and facilitates digitalization, but this method requires extremely high prediction accuracy.

2.2   Three-level DC/DC resources conversion

Like DC/AC, DC/DC level refers to the voltage class of the DC/DC output (before rectifier/filter circuit) relative to the phase voltage of the neutral point N of bus. Take the typical diode-clamped full-bridge DC/DC as an example: the existent and only existence of $\pm U/2$ belongs to two levels, and the line voltage is 0 and $\pm U$; the existent and only existence of 0 and $\pm U/2$ belongs to three levels, and the line voltage is 0, $\pm U/2$, and $\pm U$. The main circuit topology and the output voltage waveform (before filtering circuit) are as shown in Fig. 2.8. The principle of multilevel is similar to that of DC/AC; no more detailed description will be given herein. Due to the fact that all secondary rectifier and filter circuits are the same or similar, therefore, in this section, all secondary circuits of the topological structure (within the dotted line box) have been omitted.

Compared with the two-level, the three-level has the following advantages: (1) low harmonic content, and small size of filter; (2) low switching loss, with high conversion efficiency; (3) significantly decrease of output voltage leap step, and low voltage stress; (4) small size of the device, with high power density; and (5) facilitating modular parallel connection. The multi-DC/DC has lower harmonic content, but as the level number increases, the topological structure and the control systems are getting more complicated; therefore, the three-level DC/DC power module which realizes auto connection of the power motor for large capacity resources conversion and constitutes full power coverage DC to DC power electronic resources conversion is the reasonable and economical technical route.

2.2.1   Three-level DC/DC topology

The three-level DC/DC topology develops from the three-level DC/AC topology, which can be divided into three types: diode clamped type, single tube type, and push-pull type.

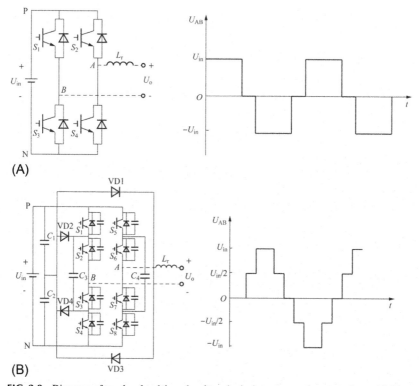

FIG. 2.8   Diagrams of two-level and three-level topological structure and output voltage. (A) Two-level and (B) three-level.

(1) *Diode clamped type*: the diode clamped type three-level DC/DC circuit consists of a prestage three-level inverter bridge arm, a voltage transformation link, and a poststage rectification link. The structure of the prestage three-level inverter bridge arm is the same as that of the diode clamped type three-level DC/AC bridge arm; therefore, the prestage circuit of this three-level DC/DC circuit can be composed of a single bridge arm or double bridge arms.

(2) *Single tube type*: The single tube type three-level DC/DC circuit consists of power elements and other auxiliary elements, which can supply power to loads directly without passing through a transformer, with power flowing unidirectionally and bidirectionally.

(3) *Push-pull type*: The push-pull type three-level DC/DC circuit consists of a prestage three-level DC/AC conversion part, a transformation link, and a poststage rectifying circuit. Each part is connected to the circuit in the push-pull mode.

Typical topological structures are as shown in Fig. 2.9. In Fig. 2.9A, Only two bridge arms use a switching tube, which is called *diode-clamped half-bridge*

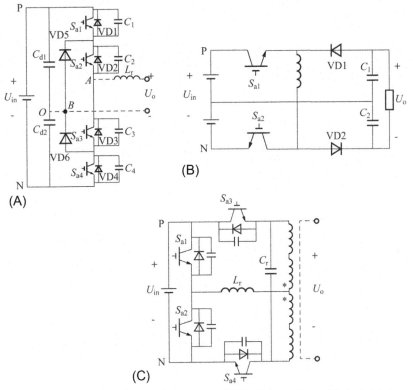

FIG. 2.9 Typical three-level topological structure. (A) Diode clamped type; (B) single tube type; and (C) push-pull type.

topology, while in Fig. 2.9B, all four bridge arms use switching tubes, which is called *diode-clamped full-bridge topology*. Besides, the deviated topologies mainly include zero voltage switch (ZVS) PWM half-bridge topology, zero voltage zero current switch (ZVZCS) PWM half-bridge topology, ZVS PWM hybrid full-bridge topology, and ZVZCS PWM hybrid full-bridge topology, which are as shown in Fig. 2.10.

2.2.2 Comparison between different diode clamped types

For the ZVS PWM half-bridge topology, the voltage stresses of all switching tubes are half of the input DC voltage; therefore, it is suitable for an application with high voltage and medium-to-large power. In this topology, the lead tube (the disconnecting time of each switching tube staggers, the first disconnected tube is called the *lead tube*, while the tube disconnected delay for a while is called the *lag tube*) can realize ZVS under a wide load range, while the lag tube can realize ZVS only with the leakage inductance. For this topology output

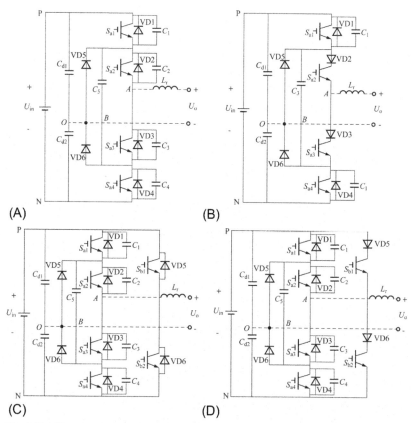

FIG. 2.10 Three-level topological structure deviated from diode clamped topological structure. (A) ZVSPWM half-bridge; (B) ZVZCSPWM half-bridge; (C) ZVSPWM hybrid full-bridge; and (D) ZVZCSPWM hybrid full-bridge.

rectifier tube, the problems of voltage spike and voltage oscillation exist due to reverse recovery.

For the ZVZCS PWM half-bridge topology, the switching tube voltage is half of the input DC voltage. It can realize lead tube ZVS under a wide range, without suffering the primary side circulation problem exists in the ZVSPWM half-bridge topology, which can effectively improve the conversion efficiency of the converter.

In the ZVS PWM hybrid full-bridge topology, one bridge arm belongs to a three-level bridge arm, with the voltage stress of the switching tube being half of the input voltage, which can realize ZVS under a wide load range; the other bridge arm belongs to a two-level bridge arm with the voltage stress of the switching tube being equal to the input voltage, which, with resonant inductance, can realize ZVS under a wide load range; the ZVS PWM output rectified waveform contains a low-frequency component, which can reduce the size of the filter.

In the ZVZCS PWM hybrid full-bridge topology, one bridge arm belongs to a three-level bridge arm with the voltage stress of the switching tube being half of the input voltage, which can realize ZVS under a wide load range; the other bridge arm belongs to a two-level bridge arm with the voltage stress of the switching tube being equal to the input voltage, which can also realize ZCS under a wide load range; the output rectified waveform contains a low-frequency component, therefore, a small output filter can be selected. The input current is close to DC current, which can reduce the size of the input filter.

Differences between the diode clamped half-bridge topology and the full-bridge topology are as follows:

(1) In terms of the circuit structure, the half-bridge structure is simple, while the full-bridge structure is complex. The voltage stress of the switching tube of both types is half of the input voltage. The diameter of the primary winding wire of the half-bridge transformer is larger than that of the full-bridge type, while the full-bridge type has more primary winding turns than the half-bridge type.

(2) In terms of control, compared with the half-bridge topology, the full-bridge topology has more switching tubes; therefore, it has more switching modes, and the control of the same is more complicated.

(3) In terms of power output, the primary voltage amplitude of the full-bridge transformer is $\pm U$, while the primary voltage amplitude of the half-bridge transformer is $\pm U/2$. Therefore, the rated output power of the full-bridge type is twice that of the half-bridge. The half-bridge type is not applicable with large power requirements.

2.2.3   Comparison between push-pull type and diode clamped type

The push-pull type topology can be further divided into the ZVS PWM push-pull type and ZVZCS PWM push-pull type.

Compared with the diode clamped type, the push-pull type has advantages of simple structure, low conduction loss, and simple drive circuit, but the scope of application is narrow, which is mainly applied to a situation with low input voltage and small-medium power.

Comparison results between push-pull type and diode clamped type are as shown in Table 2.2.

2.2.4   Comparison between single-tube type and diode clamped type

There are six single-tube type topological structures: Buck, Boost, Buck-Boost, Cuk, Sepic, and Zeta topological structures. Compared with the diode clamped type, the single-tube type is featured by electrical isolation free and simple

TABLE 2.2 Comparison between push-pull type and diode clamped type

Topological structure	Push-pull	Half-bridge	Full-bridge
Drive circuit	Simple	Medium	Complex
Loss	Small	Medium	Large
Voltage stress of switching tube	Input voltage	Half input voltage	Half input voltage
Voltage amplitude at primary side of transformer	Input voltage	Half input voltage	Input voltage
Power range	Hundreds of watt thousands of watt	Hundreds of watt—thousands of watt	Hundreds of watt—hundreds of kilowatt

structure, but it not applicable to an occasion with high input and output, especially for the last four types of converters; the voltage stress on the switching tube will be the sum of the input voltage and the output voltage. By adopting an alternating switching mode by the single-tube type, the maximum inductive current pulsation can be reduced significantly, which can widen the scope of application. In addition, the size of the energy storage element such as the inductor and the capacitor can be reduced, so as to improve the dynamic performance of the converter, and the size and weight can be reduced and the power density can be increased.

2.2.5 Improvement of half-bridge topology

The distributed resources power module adopts modular design. It is required that the DC/DC power module will be able to be applied to PV, energy storage, electric vehicle, and other applications, with small size, high power density and facilitating modular parallel connection; in addition, bidirectional power flow will be realized as the application occasion may require.

After taking all factors into consideration, the improved half-bridge three-level DC/DC topology is adopted, as shown in Fig. 2.11. This topology does not need the high-frequency isolation transformer and the secondary circuit, which has the advantages of small size, high efficiency and high power density, and the voltage stress on each switching tube is half of the input voltage. In addition, this topology adopts the structure of two half-bridges in parallel connection, which is applicable to the application occasion with high input voltage and large power.

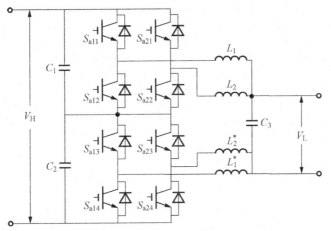

FIG. 2.11 Improvement of half-bridge topology.

2.2.6 Neutral-point potential balance control

Like the three-level DC/AC, the three-level DC/DC also suffers the problem of neutral-point potential fluctuation. During DC/DC operation, the load current flows through the neutral point and voltage on the two capacitors are unevenly divided, which results in potential fluctuation of the neutral point. The potential fluctuation of neutral point may lead to the following two problems: bring low-order harmonics to the output voltage, which will result in serious distortion of output voltage waveform and reduce the quality of output power, leading to uneven voltage on the power switching tube; or over-high voltage may cause the damage of capacitor or other elements at the DC side.

Potential unbalance of the neutral point is the main factor that holds back three-level DC/DC application, and this problem can be solved via two aspects—hardware and software.

2.2.6.1 Hardware control

The neutral point potential balance should be ensured by changing the main circuit, such as using independent DC power sources or increasing the capacitance. However, this may significantly increase the equipment costs and increase the complexity of circuit design, which is economically infeasible and impracticable. Therefore, current research mainly focuses on software control.

2.2.6.2 Software control

Based on the circuit characteristics of the three-level DC/DC, on the premise of ensuring the sum of the duty ratio of the two switching tubes remain unchanged, increase the duty ratio of one switching tube and decrease the other; capacitor voltage balance can be realized without affecting the output voltage, so as to ensure neutral point potential balance.

For the problem of neutral point potential fluctuation, the double phase-shift PWM compound control strategy can be adopted. Change the original phase-shift PWM with single phase shifting angle into a phase-shift PWM with double phase shifting angle, generating dissymmetry PWM control pulse so as to realize neutral point potential regulation by modulating dissymmetry. In addition, as the load degree increases, the regulating capacity will increase, which has sound dynamic performance.

2.3 Three-level two-way DC/AC power module

The DC/AC power module can realize DC/AC conversion. The main flow power unit is divided into three power classes (i.e., 10/20/50 kW). By adopting modular design, free combination and multiunit in parallel connection can be realized, so as to meet the different capacity requirements of different project, and realize power matching and "plug-and-play." After configuring different software functions in the module, it can be applied to PV grid connection, wind turbine grid connection, energy storage conversion, electric vehicle two-way charging, AC/DC microgrid, and other occasions. The DC/AC power module adopts the rack-mount design, with cabinet height being $2U$, $4U$, and $6U$ ($1U = 44.45$ mm), respectively, as shown in Fig. 2.12. For the modules with larger powers such as 250 and 500 kW, the standard cabinet design will be adopted.

2.3.1 Functional features

The DC/AC power module is applicable to plenty of power electronic equipments, such as energy storage converter, PV converter, electric vehicle charging (discharging) motor, wind turbine converter, and AC/DC microgrid coordination controller. Different applications can be built in the module according to the actual conditions of application occasions, so as to meet different control requirements and realize free power combination. It is applicable to new energy power generation fields such as energy storage, PV, electric vehicle, and microgrid.

FIG. 2.12 10 kW, 20 kW, and 50 kW DC/AC power module.

Characteristics of the DC/AC power module are as follows:

(1) Three-level topology and low harmonic content
(2) Low switching loss and high machine efficiency
(3) Small size and high power density
(4) Flexible configuration, facilitating free combination and free parallel connection
(5) Supporting remote power regulation

2.3.2 Topological structure

DC/AC power modules with three power classes adopt the same main circuit topology, as shown in Fig. 2.13. The main bridge arm adopts T-type diode clamped type three-level topological structure. The filter circuit adopts the output LCL filter scheme based on common-mode and differential-mode decoupling. This scheme reduces the coupling degree between the common-mode and differential-mode filter circuits and facilitates separate design of the circuit; this topological structure effectively restrains the common-mode voltage, reduces power loss, and realizes two-way energy flow and DC-AC conversion.

2.3.3 Control mode

2.3.3.1 PQ control

PQ control, also known as constant power control, refers to control over the output active power and the reactive power of the distributed resources. For this method, the reference values of the active power and the reactive power are set, and the actual output active power and the reactive power are controlled to track such reference values. PQ control is applicable to occasions with stable voltage and frequency at AC side of the system, i.e., the voltage and frequency of the system are supported by the macrogrid or other main power sources. As

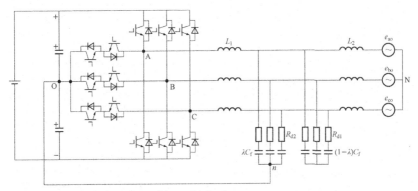

FIG. 2.13 Topological structure of the main circuit.

the voltage and the frequency change within the normal range, the output active power and reactive power from the distributed resources will remain constant.

2.3.3.2 V/f control

V/f control, also known as constant voltage/constant frequency control, refers to the control over the output voltage and the frequency of the distributed resources. For this method, the reference values of the voltage and the frequency will be set, and the actual output voltage and the frequency will be controlled to track the reference values. The objective of *V/f* control is to ensure the output voltage and frequency of the distributed resources can remain constant, and the voltage and the frequency of the distributed resources is unrelated to the output power of the distributed resources. *V/f* control is generally applicable to occasions where the distributed resources will serve as independent main power sources (able to provide voltage and frequency support).

2.3.3.3 Droop control

Droop control refers to a method of simulating the drooping characteristics of a traditional generator set. This control method is generally applicable to occasions where multiple communication line-free inverters are connected in parallel. The working principle is as follows: each inverter self-detects its own output power and then adjusts the same against the reference output voltage amplitude and the frequency obtained by carrying out droop control, so as to realize reasonable allocation of active power and reactive power of the system. This control method is also applicable to macrogrid connection. In case of voltage and frequency changes of the grid, each distributed resource is able to adjust the output active power and reactive power automatically, and participates in the regulation of grid voltage and frequency.

2.3.3.4 VSG control

VSG control, also known as *virtual synchronous generator control*, can be used not only to simulate the external droop characteristics of the traditional generator set, but also the inertia and damping characteristics of the synchronous generator, which is mainly used for the regulation of grid voltage and frequency. The working principle is as follows: the inverters detect the voltage and frequency of the grid and adjust their output active power and reactive power automatically, which can solve the low inertia problem of the traditional inverters.

2.3.3.5 V control

V control, also known as *DC voltage stability control*, is applicable to voltage stability control at the DC side, such as the voltage control at the output side of the PV generation system after DC/DC conversion, and DC bus voltage control in the microgrid system.

2.3.4 Technical indexes

For the technical indexes of the DC/AC power module, please refer to Table 2.3.

TABLE 2.3 Technical indexes of DC/AC power module

S/N	Technical indexes		Technical parameters
1	Basic parameters	Rated power	10 /20/50 kW
2		Maximum power	11/22/55 kW
3		Voltage range at DC side	650–800 V
4		Maximum allowable open-circuit voltage	1000 V
6		Number of DC Inputs	1
7	Index parameters	Maximum efficiency	96.0%/97.0%/98.0%
8		Self-energy consumption at nighttime	<10 W
9		Noise	<65 dB
10		Cooling mode	Air cooling
11	Environmental parameters	Operating ambient temperature	−25°C to +60°C
12		Relative humidity	0%–90%, condensate-free
13		Maximum allowable altitude	3000 m (when the altitude >3000 m, the rated values will be decreased and applied)
14	Structural parameters	IP grade	IP20
15		Isolation mode	Nonisolated
16		Dimension (W/H/D, mm)	10 kW: 445/89/450 20 kW: 445/177/450 50 kW: 445/266/450
17		Installation mode	Rack-mounted
18	Communication parameters	Communication interface	RS485
19		Communication protocol	Modbus

2.4   Three-level two-way DC/DC power module

The DC/DC converter power module can realize DC/DC conversion; the main flow power unit is divided into three power classes: 10/20/50 kW. In terms of application installation as well as the structure and appearance design, the DC/DC power module adopts the same design philosophy as that of the DC/AC design: by adopting modular design, free combination and multiunit in parallel connection can be realized, so as to meet the different capacity requirements of different project and realize power matching and "plug-and-play." After configuring different software functions in the module, it can be applied to energy storage, PV, electric vehicle DC microgrid, and other occasions, realizing "plug-and-play" in different application occasions. The cabinet heights are $2U$, $4U$, and $6U$ ($1U = 44.45$ mm), respectively. DC/DC module adopts the rack-mount design, while the modules with larger powers such as 250 and 500 kW adopt standard cabinet design.

2.4.1   Functional features

The DC/DC power module is applicable to plenty of power electronic equipment, such as energy storage converter, PV converter, electric vehicle off-board charging (discharging) device, and DC transformer. Free power combination can be realized according to the conditions of different application occasions, so as to be able to be applied in the new energy power generation fields such as energy storage, PV, electric vehicle, and microgrid.

Characteristics of the DC/DC power sources are as follows:

(1)   Three-level topology, with high power density
(2)   Low switching loss and high machine efficiency
(3)   Flexible configuration, facilitating free combination and free parallel connection
(4)   Supporting remote power regulation

2.4.2   Topological structure

DC/DC power modules with three power classes adopt the same main circuit topology, as shown in Fig. 2.14. In order to improve the load-bearing capacity of power elements, the parallel-connected three-level structure consisting of two I-shape non-common-grounded three-level Buck-Boost two-way converters is adopted. Then a distributed resource (such as PV element and energy storage cells) is connected, and U_H is connected to the poststage DC/AC or DC grid. According to the actual requirements of different application occasions, it is applicable to the Boost Mode and the Buck Mode, which realizes two-way energy flow and conversion of DC power in different voltage classes.

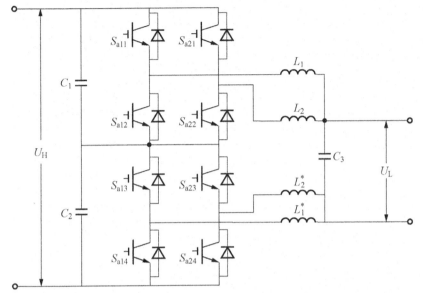

FIG. 2.14   Topological structure of main circuit of module.

2.4.3   Control mode

2.4.3.1   P control

For DC/DC, reactive power control is not required, but active power control (i.e., the P control), also known as *constant power control*, is required. This control refers to the regulation of DC/DC output active power so as to realize different power outputs. Set the reference active power, process the actual output active power of DC/DC against the reference values, and then enter PI control. The PI regulator is used to control the deviated DC/DC output power (active power) against the reference value and quickly stabilize the output power. P control of DC/DC is applicable to occasions with relatively stable system voltage at the DC side, i.e., system voltage is supported by the DC macrogrid or other DC main power sources, such as the DC energy storage system, the two-stage AC energy storage system, the DC PV generation system, and the two-stage AC PV generation system.

2.4.3.2   V control

For DC/DC, V control, also known as the *constant voltage control*, refers to the control over the DC/DC output voltage (for a two-way DC/DC, LV side output and HV side output will be distinguished). Set the reference output voltage, control the actual DC/DC output voltage against the reference value, and output PI control; as the principle of the subsequent procedures is the same as that of P control, no more details will be given herein. The V control of DC/DC converter is applicable to the following occasions:

(1) DC/DC serves as an independent power source—for example, the DC microgrid system with DC loads connected.
(2) The two-stage system consists of DC/DC and DC/AC, such as two-stage AC PV power generation system, two-stage AC energy storage system, or electric vehicle charging system. In this case, DC/DC is used to control the internal DC bus voltage; while for the two-stage AC energy storage system and the electric vehicle charging system, the DC/DC module can also be used to control voltage stability at the battery side.

2.4.3.3 I control

For DC/DC, I control, also known as the *constant current control*, refers to the control over the DC/DC output current (for two-way DC/DC, output current, or input current). Set the reference output current, control the actual DC/DC output current against the reference values, and output PI control; as the principle of the subsequent procedures is the same as that of P control, no more details will be given herein. The I control of DC/DC converter is applicable to the following occasions:

(1) Grid-connected DC system; DC/DC is used to control the current at the charging side.
(2) The two-stage system consists of DC/DC and DC/AC, such as a two-stage AC energy storage converter or an electric vehicle charging system. In this case, DC/DC is also used to control the current at the battery side.

It should be noted that the applications of P control, V control, and I control are not limited to the provided examples. For some occasions, the three controls may be required simultaneously. For example, during the constant power charging + constant voltage charging + constant current charging process of a two-stage AC energy storage system, P control, V control, and I control will be provided.

2.4.4 Technical indexes

For the technical indexes of the DC/DC power module, please refer to Table 2.4.

2.5 Power electronic resources conversion device

Different control programs should be built according to the application occasions of the distributed resources. Freely combine the three DC/AC and DC/DC power modules with different power classes as required, and install the module as per the standard mounted-type installation procedures, so as to form a product for a special field, such as the PV inverter, the wind turbine converter, and the energy storage converter, as shown in Fig. 2.15.

TABLE 2.4 Technical indexes of DC/DC power module

S/N	Technical indexes		Technical parameters
1	Basic parameters	Rated power	10/20/50 kW
2		Maximum power	11/22/55 kW
3		Voltage range at DC low-voltage side	250–800 V
4		Maximum allowable open-circuit voltage	1000 V
5		Voltage range at DC high-voltage side	650–750 V
6		Number of DC inputs	1
7	Index parameters	Maximum efficiency	98%
8		Self-energy consumption in the nighttime	<10 W
9		Noise	<65 dB
10		Cooling mode	Air cooling
11	Environmental parameters	Operating ambient temperature	−25°C to +60°C
12		Relative humidity	0%–90%, condensate-free
13		Maximum allowable altitude	3000 m (when the altitude >3000 m, the rated values will be decreased and applied)
14	Structural parameters	IP grade	IP20
15		Isolation mode	Nonisolated
16		Dimension (W/H/D, mm)	10 kW: 445/89/450 20 kW: 445/177/450 50 kW: 445/266/450
17		Installation mode	Rack-mounted
18	Communication parameters	Communication interface	RS485
19		Communication protocol	Modbus

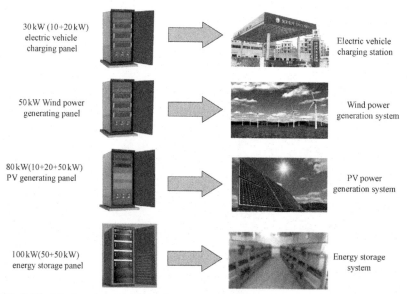

30 kW (10+20 kW) electric vehicle charging panel

Electric vehicle charging station

50 kW Wind power generating panel

Wind power generation system

80 kW(10+20+50 kW) PV generating panel

PV power generation system

100 kW(50+50 kW) energy storage panel

Energy storage system

FIG. 2.15 Distributed resources in different application fields.

The converters of PV, wind power, energy storage, and the electric vehicle will be designed to DC/DC and DC/AC. DC/DC and DC/AC are the key power electronic devices used for the connection of the distributed resources to the AC grid and to the microgrid system consisting of distributed resources, as shown in Fig. 2.16. The power converters of PV, energy storage, and electric vehicles connected to AC bus adopt a two-stage DC/DC + DC/AC combination, whereas the power converters of PV, energy storage, and electric vehicles connected to DC bus adopt DC/DC. Coordination control between AC bus and DC bus is

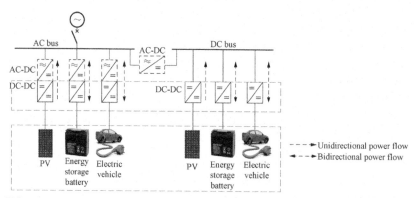

FIG. 2.16 Formation of AC&DC hybrid microgrid consisting of distributed resources via DC/AC and DC/DC.

realized through the DC/AC, realizing the DC&AC hybrid microgrid consisting of DC/AC and DC/DC modules.

2.5.1 PV converter

The PV converter is a key power electronic device within the PV generation system. The main function is to convert the DC power to AC or DC power suitable for grid connection. From the aspect of power flow direction, it belongs to a one-way power conversion device, which can be further divided into the PV AC converter and the PV DC converter according to the type of grid it will connect to.

2.5.1.1 PV AC converter

The PV AC converter is also known as the *PV inverter*. From the aspect of the power conversion stage, it can be further divided into a single-stage type and two-stage type, as shown in Fig. 2.17. The single-stage type consists of single DC/AC power module or multiple DC/AC power modules in parallel connection; the two-stage type consists of a prestage DC/DC and a poststage DC/AC in series connection, in which the prestage DC/DC consists of a single DC/DC power module or multiple DC/DC modules in parallel connection, while the poststage DC/AC consists of a single DC/AC power module or multiple DC/AC power modules in parallel connection. The main function of the PV inverter is to convert the DC power from the solar panel into AF power for AC grid connection; in addition, the PV inverter has been equipped with functions such as

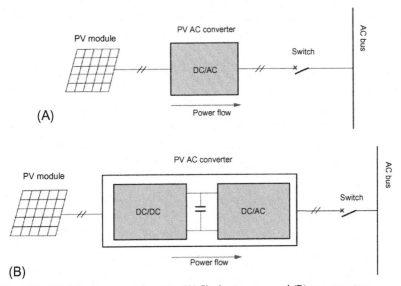

FIG. 2.17 PV AC grid-connected system. (A) Single-stage type and (B) two-stage type.

maximum power point tracking (MPPT), unit power factor grid connection, auto voltage regulation for grid connection, and power quality control, as well as AC anti-islanding protection, leak current protection, over-/under-voltage protection and over-/under-frequency protection. The PV AC converter is mainly applied to distribute AC PV, AC PV station, AC microgrid, and AC&DC hybrid microgrid.

2.5.1.2 PV DC converter

The PV DC converter consists of single DC/DC power module or multiple DC/DC power modules in flexible parallel connection, as shown in Fig. 2.18. The main function of the same is to convert the DC power from the solar panel into the DC power applicable to DC grid or DC loads. It is equipped with control functions such as MPPT, auto voltage regulation, as well as the protection functions such as DC anti-islanding, leak current protection, over-/under-voltage protection, and short-circuit protection. The PV DC converter is mainly applied to distribute DC PV, DC microgrid, and AC&DC hybrid microgrid.

2.5.2 Wind turbine converter

The wind turbine converter is a key power electronic device in the wind power generation system, which can be further divided into the wind turbine AC converter and the wind turbine DC converter, according to the type of grid it will be connected to.

2.5.2.1 Wind turbine AC converter

The wind turbine AC converter consists of a prestage AC/DC and a poststage DC/AC in series connection, realizing energy transmission via the DC link. The prestage AC/DC consists of single AC/DC power module or multiple AC/DC modules in parallel connection, while the poststage DC/AC consists of single DC/AC power modules or multiple DC/AC power modules in parallel connection. By application occasions, the wind turbine AC converter can be further

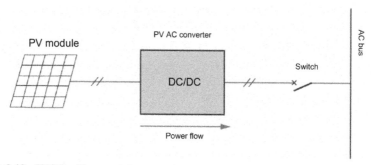

FIG. 2.18 PV DC grid-connected system.

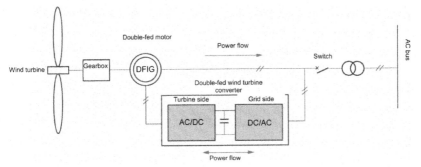

FIG. 2.19 Double-fed wind power generation system.

divided into the double-fed wind turbine converter and the direct-driven permanent-magnet synchronous wind turbine AC converter.

The double-fed wind turbine converter is a key power electronic device within the double-fed wind power generation system, as shown in Fig. 2.19. The main function is to realize variable-speed constant-frequency grid connection and power generation through excitation regulation of the rotor of the double-fed wind turbine. From the aspect of power flow direction, it belongs to a two-way power conversion device: when the wind turbine is operating at hypersynchrony speed, power flows from the rotor to the grid; when operating at subsynchronous speed, power flows from the grid to the rotor. The AC/DC converter at turbine side, supported by a pitch regulating mechanism, can realize maximum wind power capture and the function of power-factor regulation at stator side, so as to improve the power generation efficiency of the wind power generation system, while the DC/AC converter at the grid side is mainly for maintaining constant voltage on DC bus and for grid connection of unit power factor. In general, the double-fed converter is equipped with over-current protection, current leak protection, default phase protection, earth fault protection, over-/under-voltage protection, and grid failure protection.

The converter of direct-driven permanent magnet synchronous wind turbine is a key power electronic device in the direct-driven permanent-magnet synchronous AC wind power generation system, as shown in Fig. 2.20. The main function of the converter is to convert the irregular AC power generated by the wind turbine into regular AC power and feed the power to the grid. From the aspect of power flow direction, it belongs to a one-way power conversion device. The AC/DC converter at the turbine side continuously converts the AC power with variable voltage, amplitude, and frequency output from the motor rotor into DC power, and realizes stable DC voltage output from the generator under different wind speed and rotation speed conditions; while the DC/AC converter at the grid side can realize decoupling control of active power and reactive power connected to the grid by adopting the vector control method. On one hand, it maintains the voltage stability on the DC bus; on the other hand, it

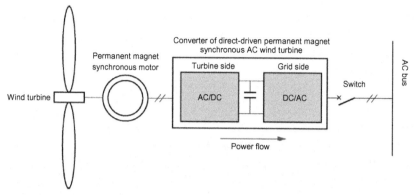

FIG. 2.20   Direct-driven permanent magnet synchronous AC wind power generation system.

feeds the DC power output from the converter at turbine side to the grid, realizing reliable grid connection of the full power converter. In general, this type of converter is equipped with overcurrent protection, current leak protection, default phase protection, earth fault protection, over-/under-voltage protection, and power failure protection.

2.5.2.2   Wind turbine DC converter

The wind turbine DC converter is a key power electronic device within the direct-driven permanent magnet synchronous DC wind power generation system, as shown in Fig. 2.21, which consists of a prestage AC/DC and a poststage DC/DC in series connection, and the DC link realizes power transmission. The prestage AC/DC consists of single AC/DC power module or multiple AC/DC modules in parallel connection, while the poststage DC/DC consists of single DC/DC power module or multiple DC/DC power modules in parallel

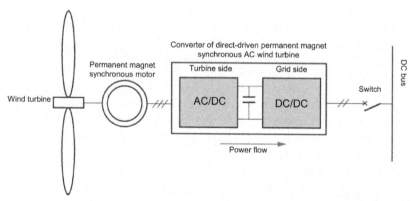

FIG. 2.21   Direct-driven permanent magnet synchronous dc wind power generation system.

connection. The main function of the converter is to convert the AC power with variable voltage, amplitude, and frequency generated by the permanent-magnet synchronous generator into DC power and feed the power to the grid. From the aspect of power flow direction, it belongs to a one-way DC power conversion device. In general, it is equipped with overcurrent protection, default phase protection, over-/under-voltage protection, and grid failure protection, and is mainly applied to the DC wind power generation system of the direct-driven synchronous generator.

2.5.3   Energy storage converter

The energy storage converter is a key power electronic device within the energy storage system, which is mainly for the power conversion between the DC power output from the energy storage device and power from the grid. From the aspect of power flow direction, it belongs to a two-way power conversion device. It can be further divided into the energy storage AC converter and the energy storage DC converter according to the type of grid it will connect to.

2.5.3.1   Energy storage AC converter

From the aspect of power conversion stage, the energy storage converter, as shown in Fig. 2.22, can be further divided into single-stage type and two-stage type. The single-stage type consists of a single DC/AC power module or multiple DC/AC power modules in parallel connection. The two-stage type consists

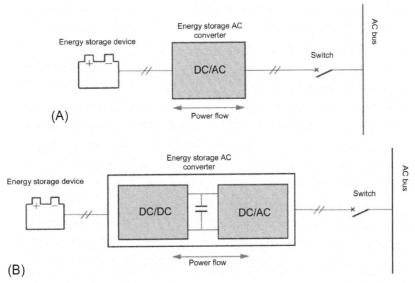

FIG. 2.22   Energy storage AC grid-connected system. (A) Single-stage type and (B) two-stage type.

of a prestage DC/DC and a poststage DC/AC in series connection, in which the prestage DC/DC consists of a single DC/DC power module or multiple DC/DC modules in parallel connection, while the poststage DC/AC consists of single DC/AC power module or multiple DC/AC power modules in parallel connection. This converter is mainly used for power conversion between chemical energy and AC energy for the energy storage device, so as to realize peak-load shifting and distributed generation output smoothing; The energy storage system has two operation modes—the grid-connected mode and the grid-disconnected mode—and is equipped with the functions of charging/discharging control, seamless grid connection and disconnection, and ultra-quick charging and discharging shifting, as well as the anti-islanding protection, leak current protection, and over-/ under-voltage protection. It is mainly applied to AC energy storage, AC microgrid, and AC/DC hybrid microgrid.

2.5.3.2 Energy storage DC converter

The energy storage DC converter consists of a single DC/DC power module or multiple DC/DC power modules in flexible parallel connection, as shown in Fig. 2.23. This converter is mainly used for power conversion between chemical energy and AC energy for the energy storage device. It has two operation modes—the grid-connected mode and the grid-disconnected mode. It is equipped with functions of self-adaption charging and discharging control, precharging, and DC bus voltage stabilization, as well as the anti-islanding protection, leak current protection, and over-/under-voltage protection. It is mainly applied to DC energy storage, the DC microgrid, and the AC/DC hybrid microgrid.

2.5.4 Electric vehicle charging/discharging machine

The electric vehicle charging/discharging machine is a key power electronic device within the electric vehicle charging and discharging system, which is mainly used for power conversion between chemical energy and electric energy for the on-board energy storage device of the electric vehicle. From the aspect of

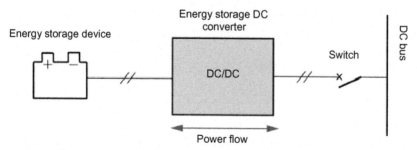

FIG. 2.23 Energy storage DC grid-connected system.

power flow direction, it belongs to a two-way power conversion device. It can be further divided into the AC charging/discharging machine and the DC charging/discharging machine, according to the type of grid it will connect to.

2.5.4.1 AC charging/discharging machine

From the aspect of the power conversion stage, the AC charging/discharging machine, as shown in Fig. 2.24, can be further divided into single-stage type and two-stage type. The single-stage type consists of a single DC/AC power module or multiple DC/AC power modules in parallel connection; the two-stage type consists of a prestage DC/DC and a poststage DC/AC in series connection, in which the prestage DC/DC consists of a single DC/DC power module or multiple DC/DC modules in parallel connection, while the poststage DC/AC consists of a single DC/AC power module or multiple DC/AC power modules in parallel connection. It is mainly used for power conversion between chemical energy and electric energy for the on-board energy storage device of the electric vehicle. Supported by the electric vehicle battery management system, adopting V2G technology, it realizes the interaction between the electric vehicle and the grid. The machine is equipped with the functions of smart charging and discharging control, ultraquick charging and discharging shift, as well as the leak current protection, over-/under-voltage protection, and over-/under-frequency protection, which is mainly applied to the electric vehicle AC charging system, AC microgrid, and AC&DC hybrid microgrid.

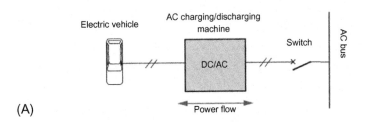

(A)

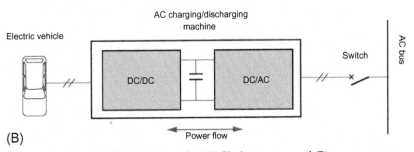

(B)

FIG. 2.24 AC charging/discharging machine. (A) Single-stage type and (B) two-stage type.

2.5.4.2 DC charging/discharging machine

The DC charging/discharging machine consists of a single DC/DC power module or multiple DC/DC power modules in flexible parallel connection, as shown in Fig. 2.25. It is mainly used for power conversion between chemical energy and AC energy for the on-board energy storage device of the electric vehicle. Supported by the electric vehicle battery management system and by adopting advanced control technologies, the interaction between the electric vehicle and the grid can be realized. The DC charging/discharging machine is equipped with the functions of smart charging and discharging control, ultraquick charging and discharging shift, as well as leak current protection, over-/under-voltage protection, and short-circuit protection, which is mainly applied to the electric vehicle DC charging system, DC microgrid, and AC&DC hybrid microgrid.

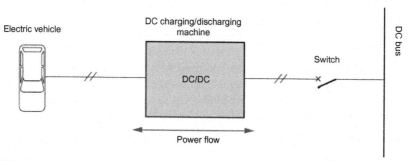

FIG. 2.25 DC charging/discharging machine.

Chapter 3

Key technologies for grid connection of distributed resources

Distributed resources feature the randomness, intermittence, and fluctuation as all electronic power resources. Thus connecting distributed resources to the grid will affect the safety and stability of the grid to a certain degree. To solve this problem, the following key technologies, such as safety grid connection of distributed resources, islanding prevention and protection, abnormal voltage and frequency auto control, seamless microgrid connection and disconnection, and active grid protection control, will be explored. This chapter gives detailed descriptions of theoretical foundations, along with simulation and experimental verification. The engineering applications of the key technologies for grid connection of distributed resources are also presented.

3.1 Low-frequency power injection type active islanding detection technology

Since most distributed resources are connected to the grid at the consumer-side. Therefore, if the output and loads are in balance locally, islanding will occur. As shown in Fig. 3.1, *islanding* refers to the condition when power supply to the grid is disconnected from the grid while distributed resources supply to loads independently and continuously, forming a self-sufficient islanding power generation system. Hazards of Islanding are as follows: it endangers the personal safety of grid maintenance personnel and consumers; it hampers the normal closing of the grid (nonsynchronous closing); the voltage and the frequency within the island are uncontrollable, which may damage distribution equipment and user equipment; islanding indicates running independently and out of monitoring of the power management department, and thus uncontrollable high hazard operations may exist.

The possibility of islanding operation is relatively high when the distributed resources are connected to the low voltage grid with a voltage class of 220 V/380 V. For the sake of safety and power quality, DG will be disconnected

Distributed Power Resources. https://doi.org/10.1016/B978-0-12-817447-0.00003-1

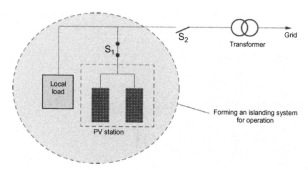

FIG. 3.1   Islanding.

from the grid rapidly and accurately in case of islanding. For distributed resources applications in the PV and energy storage AIO machine mode and in the microgrid mode, the microgrid consisting of DG and local loads has become a beneficial supplement to the macrogrid. It has two operation modes (i.e., grid-connected mode and grid-disconnected mode), which supplies power to local loads, reduces loss arising from power failure, and improves power quality and reliability. The microgrid operation control system should realize mode shifting from the grid-connection mode to grid-disconnection mode (i.e., islanding operation); therefore, the islanding detection function will be provided.

All existing international standards attach great importance to islanding detection. Standard IEEE StcL 929-2000, UL1741, and IEEE 1547 require that all grid-connected inverters should be provided with an islanding detection function, and stipulate the time limit for disconnecting the inverter from the grid when islanding is detected in case of a power outage. There are two methods for islanding detection: the passive method and the active method. For the passive method, islanding is identified on basis of electrical parameters change, such as amplitude and phase of PCC voltage, frequency, and harmonics. Main detection methods include voltage/frequency detection (VFD), harmonics detection (HD), and phase jump detection (PJD). All the provided methods are simple, and no interference will be imposed to the grid. However, detection dead zones exist, threshold setting is difficult, and it is hard to archive both dead zone reduction and maloperation reduction. For the active method, islanding is identified based on the output formula of the inverter $I = I_m\sin(2\pi ft + \theta)$. Actively impose millisecond-class intermittent disturbance on the output voltage, frequency, and power of the inverter by adding three small variables, current, amplitude, and initial phase, into the control signals of the inverter. During grid-connected operation, the voltage amplitude, phase, and frequency at PCC are restrained by the grid voltage, and signals will not be affected by such disturbance, During grid-disconnected operation, a cumulative disturbance will drive the voltage amplitude and frequency out of the normal range, then determine islanding based on the passive abnormal voltage and frequency. Active islanding detection methods include the power disturbance method, the frequency disturbance method, the phase deviation method, and the grid impedance method. Compared with the passive method, the active method has advantages of small dead zone and high detection accuracy, but disturbances are brought in, which may impose adverse impacts on inverter output, and result in power quality degradation; under different load properties, the detection results may vary, or even result in failure. When multiple inverters are operating in parallel, out-sync disturbance may greatly affect the accuracy of the detection. Research has shown that when the load quality factor is 2.5, for reliable islanding detection, a disturbance of $\pm20\%$ active power or 5% reactive power should be added. Such a huge power disturbance will seriously impact power quality. The method of loads connection/disconnection can be adopted, which has small impact on the system, but the realization and installation costs are increased, and the response speed is slow due to labor inputs.

Being able to detect islanding rapidly and accurately without degrading power quality of the grid are the basic requirements for islanding detection method. Therefore, an active islanding detection technology without depending on inverters and communication is required. This chapter introduces the application of a 20 Hz low-frequency power sources injection type active islanding detection method. With the external low-power low-frequency power module, 20 Hz component of the zero-sequence component is injected into the 380 V system, and islanding is identified according to the variation characteristics of the 20 Hz component before and after islanding occurrence. Compared with other active islanding detection method, this method does not depend on the inverter, and only one point is selected for injection, which will not degrade the power quality. In addition, there is no interference between multipoints with eliminated detecton dead zone and islanding detection is rapid and accurate, which can realize safety grid connection of the distributed resources.

3.1.1 Principle of low-frequency power injection for active islanding detection

The low-frequency power injection method for active islanding detection adopts the design idea of large generator injection stator grounding protection. For the generator injection stator grounding protection scheme, a low-power, low-frequency signal source arranged at the secondary side of the neutral grounding transformer of the generator injects low-frequency signal between the generator stator loop and the ground. Under normal operation, the signal source will not generate current or generate small current, while in case of grounding failure of the generator, stator will generate high grounding current at corresponding frequency. This method is not affected by the operating conditions, which has advantages of high accuracy, dead zone free, with excellent

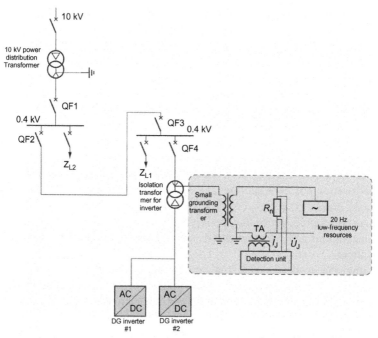

FIG. 3.2 Structure of low-frequency power injection islanding detection scheme.

performance. Based on this design idea, the structure of an islanding detection scheme is as shown in Fig. 3.2. The DG and local loads Z_{L1} are connected to the 380 V system via a dedicated line. It can be seen from the figure that QF1 is the master switch at 380 V side of the 10 kV distribution transformer, QF2 is the feeder switch of the 380 V distributed resources, QF3 is the incoming switch at resources side of 380 V user bus, QF4 is the grid-connection switch of the distributed resources, and the dashed box shows the circuit diagram of the detection scheme.

The islanding detection scheme consists of a single-phase grounding small transformer, a 20 Hz low-frequency AC power source, a detection unit, and a measuring circuit. The primary side of the single-phase grounding small transformer is connected to the neutral point of the winding at HV side of the isolation transformer of the DG inverter, and the secondary side is connected to the low-frequency AC power source.

When QF4 is closed for grid-connected operation of the distributed resources, low-frequency power source will inject a certain amount of low-frequency voltage and low-frequency current to the 380 V system via the small grounding transformer, while the detection unit will measure the voltage and current on the measuring circuit consisting of grounding transformer and low-frequency power source in a real-time manner. If islanding occurs when QF1 (or QF2 and QF3) is disconnected, the detection unit will identify islanding

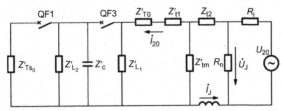

FIG. 3.3 Zero-sequence equivalent circuit of low-frequency power injection islanding detection scheme.

based on the variation characteristics of the measured low-frequency voltage and low-frequency current before and after islanding.

A 20 Hz current will be injected into the 380 V system via the winding at HV side of the isolation transformer of the inverter, which possesses zero-sequence current property. Fig. 3.3 shows the zero-sequence equivalent circuit of islanding detection scheme.

In Fig. 3.3, R_i is the internal resistance of the low-frequency power source; and R_n is the resistor connected in parallel to the secondary side of the external single-phase grounding transformer. With this resistor, the voltage $\dot{U}_J$ with 20Hz can be measured. For resistance value selection, the internal resistance R_i of the low-frequency power sources, influence imposed on power quality by the injected low voltage, and control of unbalanced current on the detection circuit should be considered, which should generally be several ohms; R_i is the internal resistance of the low-frequency power source; Z'_{t1}, Z_{t2}, and Z'_{tm} are the leakage reactance of the primary and the secondary winding and the excitation impedance of the single-phase grounding transformer, respectively; Z'_{T0} is the zero-sequence impedance of the isolation transformer; Z'_{L1} and Z'_{L2} are the zero-sequence impedances of branch loads; Z'_c is the 380 V line-to-ground capacitance; Z'_{Ts0} is the zero-sequence impedance of the 10 kV distribution transformer; and all parameters are converted to the secondary side of the single-phase grounding small transformer under 20 Hz.

Compared with other parameters, the values of the excitation impedance and the zero-sequence impedance of the load are relatively large, which have little influence on islanding detection; therefore, to facilitate analysis, the zero-sequence equivalent circuit diagram is further simplified, as shown in Fig. 3.4.

Suppose the voltage amplitude of the low-frequency power source is E; when QF1 and QF3 are closed, the 20 Hz voltage and current amplitude measured by the detection unit is as follows:

$$U_J = \frac{(Z'_{Ts0}//Z'_c + Z'_{T0})//R_n}{(Z'_{Ts0}//Z'_c + Z'_{T0})//R_n + R_i} \times E \tag{3.1}$$

$$I_J = \frac{U_J}{Z'_{Ts0}//Z'_c + Z'_{T0}} \tag{3.2}$$

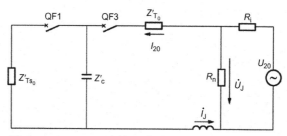

FIG. 3.4 Simplified zero-sequence equivalent circuit diagram.

Due to the 380 V system is a solidly grounded system, under normal operation, capacitive current on the cable is very small, which can be ignored. Therefore, Formula (3.1) and Formula (3.2) can be simplified as

$$U_J = \frac{(Z'_{Ts0} + Z'_{T0}) // R_n}{(Z'_{Ts0} + Z'_{T0}) // R_n + R_i} \times E \tag{3.3}$$

$$I_J = \frac{U_J}{Z'_{Ts0} + Z'_{T0}} \tag{3.4}$$

If islanding occurs when QF1 is disconnected, the 380 V system becomes an ungrounded system; in this case, the 20 Hz voltage and current amplitudes measured by the detection unit are as follows:

$$U'_J = \frac{(Z'_c + Z'_{T0}) // R_n}{(Z'_c + Z'_{T0}) // R_n + R_i} \times E \tag{3.5}$$

$$I'_J = \frac{U_J}{Z'_c + Z'_{T0}} \tag{3.6}$$

Since $Z'_c >> R_n >> Z'_{Ts0}$, with Formula (3.3) and Formula (3.5), we can see that the 20 Hz voltage U_J measured before islanding is greatly lower than the 20 Hz voltage U'_J measured before islanding; similarly, with Formula (3.4) and Formula (3.6), the 20 Hz current I_J measured before islanding is greatly higher than 20 Hz current I'_J measured after islanding.

3.1.2   Circuit parameters design and power quality impact analysis

Main designed circuit parameters include the frequency of injection signal f_d, the capacity of the low-frequency power module S_d, rated output voltage E, the no-load voltage ratio of the single-phase grounding transformer, and the resistor R_n connected in parallel at the secondary side of the transformer. These parameters are correlative with each other. Comprehensive consideration will be performed by combining the factors such as sensitivity, rapidity, and impact on power quality of the grid.

3.1.2.1 Injected signal frequency selection

For injected signal frequency selection, the frequency will be lower than the 50 Hz system frequency, and no frequency will be generated automatically by the grid, so as to eliminate the influence imposed on the measuring circuit by the 50 Hz voltage current and other higher harmonic components of the grid.

On one hand, frequency f_d injection signal determines the islanding detection speed. For the calculations of low-frequency current and voltage, real-time sampling data in a signal circle T_d is required $T_d = l/f_d$. From the aspect of protection sampling time, the lower the frequency is, the longer the protection sampling time will take, which will affect the speed of islanding detection.

On the other hand, in Fig. 3.4, all the zero-sequence impedance Z'_{T0} of the isolation transformer of the inverter, the zero-sequence impedance Z'_{Ts0} of the 10 kV distribution transformer, and the 380 V line-to-ground capacitive reactance Z'_c are related the injection frequency f_d, and $Z'_c \gg Z'_{Ts0}$ is required. Low frequency will result in small zero-sequence impedances Z'_{T0} and Z'_{Ts0} of the transformer and large line-to-ground capacitive reactance Z'_c, which can reduce the voltage drop generated by the low-frequency current on the circuit and reduce current diffusion of the low-frequency current via ground capacitance. When the power injected power is determined, the injection efficiency and the islanding detection sensitivity can be increased.

From the aspect of power quality, as shown in Fig. 3.4, under normal operation, the low-frequency zero-sequence component U_{20} of the low-frequency component injected from the neutral point of the isolation transformer lead from QF1 is as follows:

$$U_{20} = \frac{Z'_{Ts0}//Z'_c}{Z'_{Ts0}//Z'_c + Z'_{T0}} \times U_J = \kappa E \tag{3.7}$$

where, κ stands for voltage gain factor,

$$\kappa = \frac{Z'_{Ts0}//Z'_c}{Z'_{Ts0}//Z'_c + Z'_{T0}} \times \frac{(Z'_{Ts0}//Z'_c + Z'_{T0})//R_n}{(Z'_{Ts0}//Z'_c + Z'_{T0})//R_n + R_i} \tag{3.8}$$

According to Formula (3.8), selection of f_d will impact power quality. Since $Z'_c \gg Z'_{Ts0}$, we can obtain

$$\frac{1}{j2\pi f_d C'} \gg j2\pi f_d X'_{Ts0} \tag{3.9}$$

In the Formula, C' stands for the 380 V line-to-ground capacitance, X'_{Ts0} stands for the zero-sequence reactance of the 10 kV distribution transformer, then

$$f_d \ll \frac{1}{2\pi\sqrt{C'X'_{Ts0}}} \tag{3.10}$$

Let $f_r = \frac{1}{2\pi\sqrt{C'X'_{Ts0}}}$ be the natural frequency of the system. Based on empirical values, suppose $C' = l\mu F$, and $X'_{Ts0} = 0.2\,\Omega$, then f_r is approximately equal to 360 Hz. Due to the injected frequency $f_d \leq f_r$; thus, it has $f_{rd} = \frac{f_r}{10}$.

It will be noted that the principle of external 20 Hz power source injection has been applied to the rotor grounding protection for the large-scale generator, and compared with the 380 V system, the application conditions of the same are worse. Reliability of the 20 Hz power module and measurement and calculation methods for low-frequency current and voltage have been verified in the engineering practice, which can be used for reference. This is the reason for selecting 20 Hz as the frequency of the injection power.

3.1.2.2 Low-frequency power module parameter selection

The 20 Hz component injected to the 380 V system possesses zero-sequence current property and flows evenly in Phases A, B, C, which, in principle, will impact the power quality, and the extent of the impact depends on the amount of 20 Hz component injected. Parameters of low-frequency power module include capacity S_d, internal resistance R_i, rated output voltage E, and so forth, which will be considered in combination with the no-load voltage ratio of the single-phase grounding transformer and the resistor R_n connected in parallel at the secondary side of the transformer.

Capacity selection for the external low-frequency power module will be based on the principle of ensuring that the measuring accuracy of the 20 Hz voltage and current on the detection circuit meet the requirements for identification, and the influence on power quality will be reduced as much as possible. Under normal operation, considering the unbalanced voltage of the neutral point of the isolation transformer of the inverter is small, therefore, no high power frequency current will flow through the detection unit TA, as shown in Fig. 3.2; thus TA saturation is impossible. Therefore, a Class 0.2S high-accuracy current converter can be selected as the TA, so as to realize milliampere-class accurate measurement, and facilitate reduction of 20 Hz injection component. From the aspect of power quality, under normal operation, the supply voltage deviation caused by the low-frequency zero-sequence voltage injected for islanding detection is as follows:

$$\eta = \frac{U_{20}}{U_n} \tag{3.11}$$

where U_n stands for the rated voltage of the 380 V system. GB/T 12325-2008, the *Power Quality—Deviation of Supply Voltage*, stipulates that the deviation of supply voltage will not exceed $\pm 7\%$. Therefore, the rated output voltage of the low-frequency power module is as follows:

$$E < \frac{\eta U_n}{\kappa(f_d)} \tag{3.12}$$

Besides, the total harmonic distortion (THD) caused by the 20 Hz component injection can be expressed as follows:

$$\lambda = \frac{U_{20}}{\sqrt{U_n^2 + U_{20}^2}} \tag{3.13}$$

Without considering the background harmonics, according to GB/T 14549-1993, the *Quality of Electric Energy Supply—Harmonics in Public Supply Network*, the total harmonic distortion will be less than 5%. Thus,

$$U_{20} \leq \frac{\lambda}{\sqrt{1 - \lambda^2}} U_n \tag{3.14}$$

Therefore, amplitude of the rated output voltage of the low-frequency power module will conform to the following:

$$E \leq \frac{\lambda}{\kappa \sqrt{1 - \lambda^2}} U_n \tag{3.15}$$

Selections of the no-load voltage ratio of the single-phase grounding transformer and the resistor R_n will also depend on the sensitivity and the power quality of islanding detection, besides the interference imposed on islanding detection by the power frequency unbalance current on the detection circuit caused by unbalance voltage of the neutral point of the isolation transformer.

As mentioned previously, $Z'_c \gg R_n \gg Z'_{Ts0}$; therefore R_n can be the geometrical mean of Z'_c and Z'_{Ts0}.

$$R_n = \sqrt{Z'_c Z'_{Ts0}} \tag{3.16}$$

Besides, from Formula (3.8), we can obtain

$$\kappa \approx \frac{R_n Z'_c}{(Z'_c + Z'_{T0} + R_i) R_n + R_i (Z'_c + Z'_{T0})} \tag{3.17}$$

As we can see, R_n is relatively large and κ is relatively small—that is, when the injection frequency f_d and the rated output voltage E are determined, a larger R_n may reduce the impact on power quality, restrain power frequency unbalance current on the detection circuit, and improve the islanding detection accuracy.

3.1.3 Detection criteria

For island identification, the identification method of absolute amplitude comparison is adopted. According to the variation characteristics of the 20 Hz measured voltage U_J before islanding and the 20 Hz measured current I_J, the action formula is as follows by adopting the principle of absolute amplitude comparison:

$$\left| \dot{U}_J \times \dot{I}'_J \right| > k \left| \dot{U}'_J \times \dot{I}_J \right| \tag{3.18}$$

In Formula (3.18), $\dot{U}_J$ is the value measured at present by the voltage measuring branch, $\dot{U}'_J$ is the value measured before at ΔT by the voltage measuring branch; $\dot{I}_J$ is the value measured at present by the current measuring branch; $\dot{I}'_J$ is the value measured before at by the current measuring branch, and κ is the braking coefficient, $\kappa \gg 1$. When the QF1 as shown in Fig. 3.2 is closed for the grid-connected operation of the distributed resources, 20 Hz measured voltage $\dot{U}_J, \dot{U}'_J$ and the measured current $\dot{I}_J, \dot{I}'_J$ are basically the same, with actuating value $\left|\dot{U}_J \times \dot{I}'_J\right|$ significantly less than the braking value $k\left|\dot{U}'_J \times \dot{I}_J\right|$. If QF1 is disconnected, according to the previous analysis conclusion, the action formula $\left|\dot{U}_J \times \dot{I}'_J\right| > k\left|\dot{U}'_J \times \dot{I}_J\right|$ is satisfied, indicating the distributed resources is operated in the islanding mode. In the formula, the external 20 Hz power source injection stator grounding protection for the generator is used as reference for the calculation of the 20 Hz low-frequency component, and Fourier algorithm is adopted. Take 10 Hz as the reference frequency; the secondary harmonic component of the 10 Hz component is extracted for the calculations of 20 Hz voltage and current. With this method, the 20 Hz component can be extracted effectively, and other interference signals with 10 Hz multiplied frequency (including the 50 Hz power frequency) can be filtered. Fig. 3.5 shows the characteristics of islanding detection action, as well as the identification procedure.

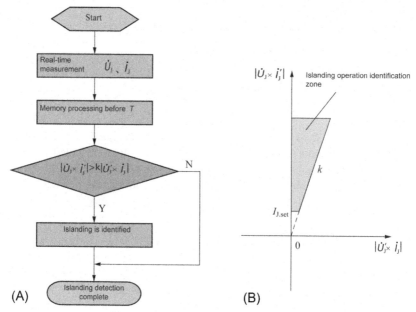

(A)   (B)

FIG. 3.5   Active islanding detection criteria. (A) Action characteristics and (B) procedures.

3.1.4 Simulation verification

The simulation model system is established based on Figs. 3.2 and 3.3 under Matlab/Simulink environment, as shown in Fig. 3.6.

According to the analysis in Section 3.1.2, the element parameters for simulation verification are as follows: voltage of the three-phase grid is 380 V/50H, capacity S_d of the 20 Hz power module is 30 W, output voltage E is 15 V, internal resistance of the low-frequency power sources is 30 Ω; the resistor R_n connected in parallel at the secondary side of the single-phase grounding transformer is 6 Ω, no-load voltage ratio of the single-phase grounding small transformer is 220/36, rated power of the inverter is 10 kW, active power of the RLC load in parallel connection is 10 kW, and the reactive power is 80 var.

Fig. 3.7 shows the simulation results. During grid-connected operation, the zero-sequence impedance of the 10 kV distribution transformer is Z'_{Ts0}, which is far smaller than the resistor R_n in parallel connection. The 20 Hz measuring voltage is small, and after boosting via the single-phase grounding small transformer, the measuring voltage is still less than 2% of the rated phase voltage; while the 20 Hz measuring current is relatively large at about 0.5 A; therefore the 20 Hz low-frequency component injected into the 380 V system hardly affect the power quality, and the system voltage and current waveforms are good. QF1 is disconnected at 0.1 s and the distributed resources and the local loads are operated under the islanding mode. Due to the zero-sequence, impedance of the 10 kV distribution transformer is disconnected from the grid and the 20 Hz measuring voltage increases in the islanding mode, which reaches up to about 2.5 V. The 20 Hz measuring current is the capacitance current, which is small. The simulation results verify the theoretical analysis results in Section 3.1.1.

3.1.5 Engineering test

Fig. 3.8 shows the interface device for safe grid connection of the distributed resources. Low-frequency power source injected active islanding detection injection circuit and realization criteria are integrated in the grid-connection interface device for the distributed resources. This grid-connection interface device integrates functions of measurement, protection, communication, connecting control, switchgear, and low-frequency power source, which is equipped with active islanding detection function. The grid-connection interface device realize integration and simplification of the PCC equipment of the distributed resources, which conform to the applicable national standards and specifications in relation to grid connection of distributed resources, as well as the requirements for the grid-connected device of the distributed resources.

The installed capacity of the distributed PV generation is 15 kW, which is connected to the small roof distributed PV PCC on the 380 V power system.

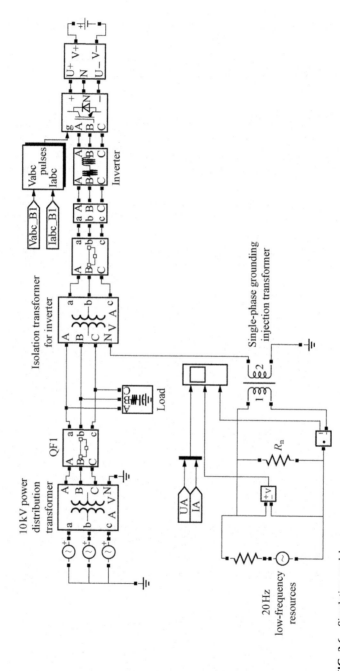

FIG. 3.6 Simulation model.

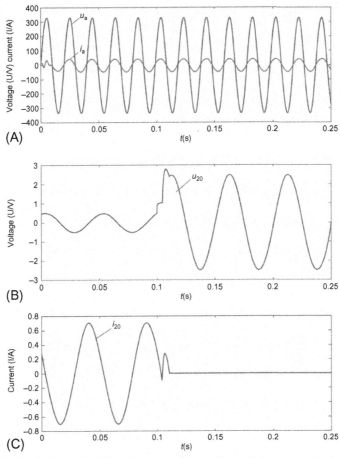

FIG. 3.7 Simulation result. (A) 380 V system voltage and current; (B) measured voltage under 20 Hz; and (C) measured current under 20 Hz.

20 Hz power resource, with power of 30 W

FIG. 3.8 Grid connection interface device.

TABLE 3.1 Comparison of phase A THD before and after injection

Injection status	Phase A voltage THD (%)	Phase A current THD (%)
Before injection	3.168	4.891
After injection	3.484	5.241

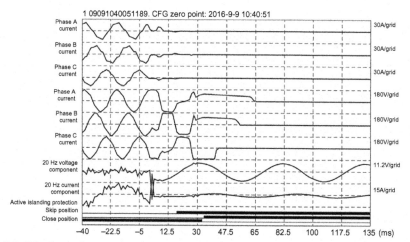

FIG. 3.9 Recording waveform for action of low-frequency power injected type active islanding detection.

This small-scale PV station adopts the grid connection interface device as shown in Fig. 3.8 for grid connection. The rated power of the inverter is 15 kW. In the experiment, the PV DC output voltage is 499.75 V, DC input current is 19.99A, and the DC side input power is about 10 kW. Differences between the voltage THD as well as current THD before and after low-frequency signal injection are small, as shown in Table 3.1, which indicates that the impact on power quality can be ignored, for the volume of injected low-frequency signal is small.

Fig. 3.9 shows the recording waveform for the action of the grid-connection interface device at PCC. Due to the injected 20 Hz component being small, for the convenience of demonstration, the scale of the 20 Hz measured voltage $\dot{U}_J$ and the measured current $\dot{I}_J$ have been zoomed up. During the experiment, when disconnecting the upper-stage 380 V breaker (equivalent to the QF3 in Fig. 3.2) at PCC of the distributed PV generation, the device will identify islanding and send the trip instruction in 20ms, and the breaker at PCC of distributed PV generation (equivalent to QF4 in Fig. 3.2) will strip in about 20ms, shifting from the closing position to the trip position and disconnect the grid-connected inverter from the grid.

3.2   Auto overvoltage/power (U/P) control technology

For grid connection of distributed resources to the grid, in order to increase the capacity penetration, the ratio of the total connected capacity of the distributed resources and the total loads of the system is defined (i.e., the technical requirements for capacity penetration). The complex structure of the grid and great amount of the connecting nodes result in low capacity penetration for grid connection of distributed resources. Therefore, the distributed resources will be adjusted appropriately according to the actual conditions of the grid, so as to increase the capacity penetration for grid connection. After distributed resources is connected to the grids, the flow distribution of the traditional grids will change, or even change to a reserve direction. Reasonable distribution of the distributed resources may improve the steady voltage at PCC. Unrestrained large-scale connection of distributed resources to the grid and loads variability may result in severe voltage deviation, fluctuation, and even result in overvoltage, which may further result in inverter disconnection due to overvoltage; In this case, the normal and safe grid-connection of the distributed resources cannot be guaranteed, and the capacity penetration of the distributed resources will be reduced. This section introduces the overvoltage/power (U/P) auto control technology, and solves the overvoltage problem caused by over active power output from the distributed resources.

3.2.1   Analysis of overvoltage during grid connection of distributed resources

3.2.1.1   Analysis of overvoltage during grid connection of distributed resources

Fig. 3.10 shows the typical voltage distribution after multiple PV DGs are connected to the grid. For the convenience of analysis, the voltage class of the grid is considered to be low and the circuit is short, the line-to-ground capacitance and other factors other than the self-impedance of the line are ignored, and the initial voltage of the line is the rated voltage U_N, in which there are totally n loads are connected to the line, and $R_k + jQ_k$ stands for

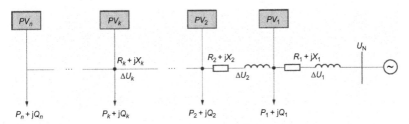

FIG. 3.10   Typical voltage distribution after grid connection of multiple PV DGs.

active power and reactive power of the load number k. $R_k + jX_k$ stands for the impedance of the feeder of Section K. PV_k stands for the PV DG number k, with a capacity of P_{vk}. ΔU_k stands for the voltage drop on the impedance of the feeder of Section K.

Assume n PV DGs are connected to the grid, for the reactance X_k of the line is small, the influence of reactive power can be ignored; in this case, the expression of voltage drop is as shown in Formula (3.19).

$$\Delta U_k = U_k - U_{k-1} = -\frac{R_k \sum_{j=k}^{n} (p_j - p_{vj})}{U_{k-1}} \tag{3.19}$$

The voltage U_k at PCC PV_k is

$$U_k = U_N - \sum_{i=1}^{k} \left[\frac{R_i \sum_{j=i}^{n} (P_j - P_{vj})}{U_{i-1}} \right] \tag{3.20}$$

According to Formula (3.20), R is related to the line impedance, PV DG output, PCC position, and the initial voltage U_N of the line. In Formula (3.20), if $\sum_{j=k}^{n} P_{Vj} > \sum_{j=k}^{n} P_j$ (i.e., when the total capacity of PV DG is larger than the total power of loads), then $U_k > U_N$. If the PV capacity is larger, U_k will fall beyond the allowable upper limit of voltage deviation U_{max} of the grid, which will result in inserter trip due to abnormal voltage, and further reduce the capacity penetration for grid connection of PV DG.

3.2.1.2 Analysis of overvoltage during grid-disconnected operation of PV DG

Fig. 3.11 shows the equivalent circuit of the grid-disconnected operation of PV DG, where $E \angle \delta$ is the open circuit voltage of the inverter, $\dot{I}$ is the output current of the inverter, and $U_n \angle 0$ is the voltage of the main power source of the microgrid. For low-voltage microgrid, the resistance on the line is large, thus the inductive reactance X can be ignored.

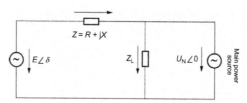

FIG. 3.11   Equivalent circuit of grid-disconnected operation of PV DG.

According to Fig. 3.11, the expression of the output current $\dot{I}$ of the inverter is as follows:

$$\dot{I} = \frac{E\angle\delta - U_N\angle 0}{R} = \frac{E}{R}\angle\delta - \frac{U}{R} \tag{3.21}$$

and the output complex power of the inverter is $\overline{S}$ as follows:

$$\overline{S} = \dot{U}_N \dot{I}^* = P + jQ \tag{3.22}$$

Combining Formula (3.21) and Formula (3.22), expressions of the active power P and the reactive power Q are as follows:

$$P = \frac{U_N}{R}(E - U_N) \tag{3.23}$$

$$Q = \frac{U_N}{R}(-\delta E) \tag{3.24}$$

It can be seen from Formula (3.23), voltage E increases as P increases.

Similar to the principle of grid connection of PV DG, considering local loads exist during grid-disconnected operation, and ignoring the influence of reactive power, the PV DG voltage can be analyzed while the actual flow distribution is as shown in Fig. 3.12.

In Fig. 3.12, assume that the PV inverter is operating with unit power factor, and only active power P_{pv} is generated. The reactive power instruction P_{ref} calculated in a real-time manner by the MPPT controller, then it has

$$P_{pv} = P_{load} + P \tag{3.25}$$

Relation between the PV output P_{pv}, load P_{load}, and voltage U_{pcc} at PCC is as shown in Table 3.2.

(1) When $P_{pv} < P_{load}$, the PV DG and the main power source supply to the loads together, and $U_{pcc} < U_N$. Under this condition, no overvoltage occurs and a certain degree of improvement will be achieved.

(2) When $P_{pv} > P_{load}$, PV generation is higher than loads consumption; thus $U_{pcc} > U_N$. Under this condition, the PCC voltage is higher than the rated voltage. In case of a sudden increase of PV generation or sudden decrease of loads which result in $P_{pv} > P_{load}$, overvoltage may occur.

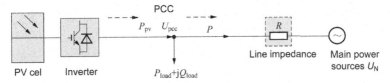

FIG. 3.12 Microgrid with PV DG connected.

TABLE 3.2 PCC voltage change caused by PV output and loads

P_{pv} to P_{load} relation	U_{pcc} changing conditions
$P_{pv} < P$ load	$U_{pcc} = U_N - \dfrac{\|P\|R}{U_{pcc}}$
$P_{pv} > P$ load	$U_{pcc} = U_N + \dfrac{\|P\|R}{U_{pcc}}$ When P_{pv} remains unchanged while $P_{load}\downarrow$, then $P\uparrow$ and $U_{pcc}\uparrow$ When P_{pv} remains unchanged while $P_{pv}\uparrow$, then $P\uparrow$ and $U_{pcc}\uparrow$

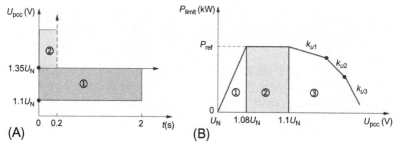

FIG. 3.13   Control line and section of U/P droop regulation (A) U/P regulation section and (B) U/P regulation line.

3.2.2 Overvoltage/power (U/P) control

For the overvoltage problem during grid-connected operation and grid-disconnected operation of PV DG, U/P droop regulation by PV inverter is adopted.

Fig. 3.13 shows the U/P droop regulation zone and the control fold, P_{limit} stands for the reference power limit, P_{ref} stands for the reference power, U_{pcc} stands for the PCC voltage of PV DG, and k_u stands for the droop regulation coefficient.

In Fig. 3.13A, the X-axis stands for time t, Y-axis stands for U_{pcc}, and both ① and ② are the time sections for P/U droop regulation.

In Fig. 3.13B, the X-axis stands for U_{pcc}, the Y-axis stands for P_{limit}, section ① indicates when MPPT tracking is in progress, section ② indicates when MPPT tracking stops, and section ③ indicates U/P droop regulation by step.

The relation between U_{pcc} and P_{limit} satisfies Formula (3.26).

$$\begin{cases} P_{limit} = P_{ref}, U_{pcc} \leq 1.08U_n\,(1) \\ P_{limit} = P_{ref}, 1.08U_n < U_{pcc} < 1.1U_n\,(2) \\ P_{limit} = P_{ref} - |k_u|\Delta U, U_{pcc} \geq 1.1U_n\,(3) \end{cases} \tag{3.26}$$

Where, $\Delta U = U_{pcc} - 1.1*U_n$, k_u stands for the droop regulation coefficient. Selection of k_u value will be as follows: overlarge k_u will result in significant change to PV DG output, which will lead to severe energy loss, and the system stability will also be affected. Small k_u will result in slow regulation speed; in this case, the inverter might not be able to regulate the voltage and frequency to the normal range within the abnormal voltage response time stipulated in the standards, and lead to the shutdown of inverter protection.

Fig. 3.14 shows the flow chart of U/P regulation, "Flag _CallMppt" stands for the MPPT tracking module call sign, and "Number" stands for the number of overvoltage.

3.2.3 Simulation verification

With the MATLAB/Simulink simulation platform, a simulation model of the single-phase grid-connected inserter based on the diagram of the grid is established, as shown in Fig. 3.10.

3.2.3.1 Calculation parameters

Simulation parameter settings are as shown in Table 3.3.

The main circuit simulation model is as shown in Fig. 3.15.

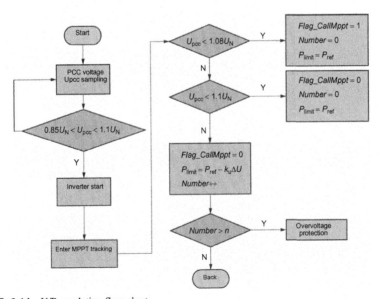

FIG. 3.14 U/P regulation flow chart.

TABLE 3.3 Simulation parameter list

Parameters	PV module						Grid parameters			
	Open-circuit voltage	Short-circuit current	Working voltage	Working current	Series connection count	Parallel connection count	Rated voltage	Line resistance	Load
Value	22V	5.3A	17.5V	4.9A	12	5	230V	1.5 Ω	5kW

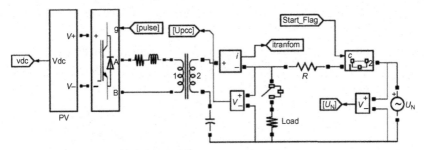

FIG. 3.15 Main circuit simulation model.

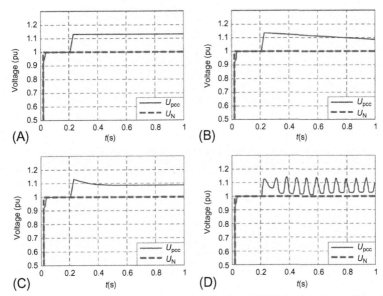

FIG. 3.16 Waveforms of PCC voltage U_{pcc} and rated voltage. (A) No-load regulation; (B) on-load regulation, with $k_u = 0.05$; (C) on-load regulation, with $k_u = 0.2$; and (D) on-load regulation, with $k_u = 2$.

3.2.3.2 Simulation result

Simulation analysis based on the U/P regulation method is carried out by Formula (3.26), and the waveforms are as shown in Fig. 3.16.

When the maximum power P_{pv} of PV DG is lower than load power Pl_{load}, simulation procedures are as follows:

(1) (1) 0–0.2 s: connect loads; in this case, $U_{pcc} < U_N$.

(2) 0.2–1 s: disconnect loads; in this case, $U_{pcc} > 1.1U_N$ and overvoltage occurs. Fig. 3.16A shows the no-load regulation, U_{pcc} exceeds allowable voltage limit; Fig. 3.16B–D show the on-load regulation, with droop coefficient k_u being 0.05, 0.2, and 2, respectively, and the regulation time is 0.36, 0.15, and 0.02 s, respectively.

According to the simulation results: the U/P droop regulation method is feasible and effective. The larger the droop coefficient is, the shorter regulation time will take (i.e., the faster the regulation speed will be); however overlarge k_u will result in system oscillation.

3.2.4   Experimental verification

A PV inverter prototype test platform is established. A PV simulator is provided at the DC side; Load 1 is for power consumption, and Load 2 is for load connection/disconnection test and for overvoltage simulation, as shown in Fig. 3.17. Main experimental parameters are set out in Table 3.4.

3.2.4.1   Test condition 1

Through connecting K1 and K2, and disconnecting K2 after the inverter runs normally for a while, the experimental waveform is as shown in Fig. 3.18.

As can be seen from Fig. 3.18, after Load 2 is disconnected, $U_{pcc} = 256$ V, overvoltage occurs, and the inverter performs droop regulation. After about 1 s, $U_{pcc} = 251$ V, which falls within the normal range, and the inverter will operate under the grid-connected mode.

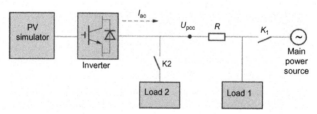

FIG. 3.17   Overvoltage test.

TABLE 3.4 Main experimental parameters

Items	Parameter values	Items	Parameter values
Maximum power at PV side	6.6 kW	Load 1	10 kW
Simulated circuit resistance R	1 Ω, 1 kW	Load 2	9 kW
Main supply voltage	230 V		

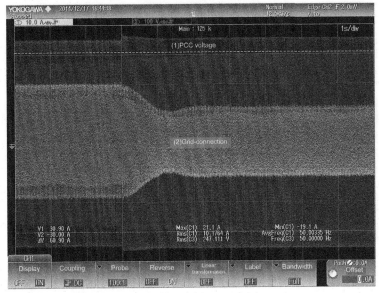

FIG. 3.18 Grid-connected current and PCC voltage waveform.

3.2.4.2 Test condition 2

Through connecting K1 and K2, and disconnecting K2 after the inverter runs normally for a while, then connecting K2 after about 40 s, and the grid-connected current waveform is as shown in Fig. 3.19.

According to Fig. 3.19, when Load 2 is reconnected, the inverter performs U/P droop regulation and *MPPT* tracking resume, and then the maximum power is reached in about 25 s.

3.3 Overfrequency/power (f/P) auto control technology

For grid connection of distributed resources, P/Q control mode is adopted. The frequency refers to the grid frequency. Because the total inertia of the synchronous generator connected to the grid is large, the frequency change is slight since the system frequency change is regulated by the grid (except for distributed resources possess characteristics of the virtual synchronous generator).

For the grid-disconnected operation of the microgrid, the power balance on the grid-disconnected microgrid will be maintained in a real-time manner. The feature of DG are multigeneration with maximum capacity. DG adopts the maximum power point tracking technology, and converts the DC power into AC power to the largest extent. When the battery is fully charged (i.e., under the high state of charge), excess energy can be neither stored nor consumed

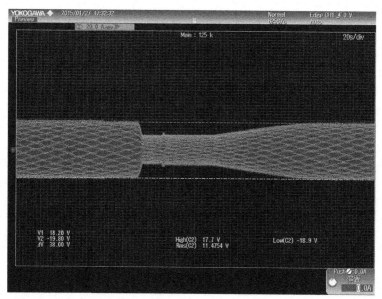

FIG. 3.19 Grid-connected current waveform.

by loads. In this case, the excess energy will result in overvoltage and over frequency, which may break the power balance and result in microgrid failure. Under this condition, the auto overfrequency and power (f/P) control technology will be adopted to properly control the DG output, so as to ensure energy balance on the grid under the grid-disconnected operation mode.

According to the power quality requirements stipulated in GB/T 15945-1995, the *Allowable Frequency Deviation of Power System*, the allowable frequency deviation of a normal grid is $\pm 0.2\,\mathrm{Hz}$. The lower limit of DG frequency is set as $f_{s1} = 50.2\,\mathrm{Hz}$, and the upper limit of DG frequency is set as $f_{s2} = 50.3\,\mathrm{Hz}$. For the grid-disconnected operation of the microgrid, DG output will be controlled as per the f/P line, as shown in Fig. 3.20, and the common P/Q control method will be adopted.

(1) The maximum frequency tracking operation mode: $f < f_{s1}$, power amount generated by DG is appropriate which will not result in overvoltage; MPPT function keeps running.

(2) Operated with the current power: $f_{s1} < f \leq f_{s2}$. Power generated by DG exceeds the load consumption but does not exceed the capacity of the main energy storage, which will not result in over-voltage. In this stage, Line 1 indicates the DG operation condition and MPPT function will be stopped, so as not to exceed the power P_0 at f_{s1}. f/P straight lines will be used to for operation control.

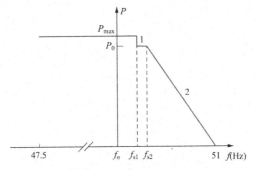

FIG. 3.20 Distributed resources f/P line control.

(3) Operated with restrained power; It has $f_{s2} < f \leq f_{max}$, and power generated by DG exceeds load consumption and is beyond the regulation capacity of the main energy storage. If the power output of DG is not restrained, overvoltage will occur. In this stage, Line 2 indicates the DG operation condition, MPPT function will be stopped, and the f/P droop control operation will be adopted. The DG inverter will regulate the power output as per Formula (3.27).

$$\begin{cases} P = P_0 + m\Delta f \\ \Delta f = f - f_{s2} \end{cases} \tag{3.27}$$

where m stands for the slope of the line.

(4) Stop power output: It has $f > 51\,\text{Hz}$, power generated by DG exceeds load consumption, SOC is high, and exceeds the regulation capacity of the main energy storage. The continuous power output of the distributed resources will result in overvoltage and further lead to system failure. In this stage, the distributed resources will stop power output (i.e., the common islanding protection for the distributed resources).

3.4 Presynchronous grid-connection control technology

In the grid-disconnected operation mode, when the power sources of the grid recover to the normal state, or when resume they from the planned islanding, the microgrid will be reconnected to the grid for operation. Phase angle difference and the frequency differences exist between the grid-disconnected voltage and the grid-connected voltage. If the distributed resources are connected to the grid directly without subjecting to synchronous control, the small voltage difference and small phase difference will impose on small connecting impedance, which will generate large surge current. The large connection surge current will trigger the overcurrent protection of the main energy storage, which will result in equipment shutdown, and the whole microgrid will black out. Therefore, before reconnecting to the grid, It is necessary to check and confirm the angle is smaller than the setting value, Which can reduce the closing surge generated at the moment of reconnection, and realize smooth switching.

3.4.1 Principle of amplitude and phase presynchronization

Assume the three-phase system voltage is

$$\begin{bmatrix} u_{a(t)} \\ u_{b(t)} \\ u_{c(t)} \end{bmatrix} = U* \begin{bmatrix} \cos(\omega t) \\ \cos(\omega t - 2\pi/3) \\ \cos(\omega t + 2\pi/3) \end{bmatrix} \tag{3.28}$$

where $\omega = 2\pi f = 100\pi$, $f = 50\,\text{Hz}$, ω stands for the angular frequency under power frequency of the grid, and U stands for the voltage amplitude.

Phase data are detected with the technologies of Software PLL (SPLL) and synchronously rotating coordinates conversion, which has sound anti-interference performance under the conditions of waveform distortion and phase change. In Fig. 3.21, after dq coordinates conversion for the three-phase voltage U_{ABC}, it is obtained that

$$\begin{bmatrix} u_d \\ u_q \end{bmatrix} = U* \begin{bmatrix} \cos(\omega t - \theta) \\ \sin(\omega t - \theta) \end{bmatrix} \tag{3.29}$$

In Formula (3.29), U stands for the three-phase voltage amplitude, and θ stands for the SPLL output. Under the unlock condition, the d-axis and g-axis components are the AC components, while under the lock condition, $\omega t = \theta$, $u_d = U$, $u_q = 0$. Therefore, phase lock can be realized by controlling a q-axis component to 0, and the information of grid voltage phase ωt can be obtained.

For presynchronous grid connection, voltages on both sides of the system will satisfy the following three conditions: same amplitude, same frequency, and same phase. Real-time synchronization of the phase means that the frequency has been locked.

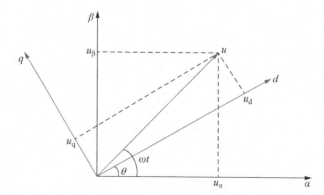

FIG. 3.21 Voltage vector diagram in αβ coordinates and dq coordinates.

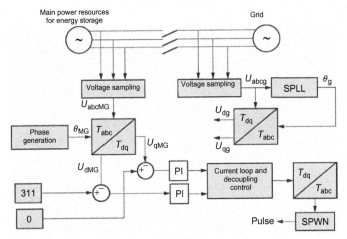

FIG. 3.22 Block diagram of V/f control under grid-disconnected operation.

Therefore, to realize synchronization of voltage on both sides, the voltage amplitude and phase at both sides will be synchronized. Fig. 3.22 shows the block diagram of V/f control of the main energy storage under the grid-disconnected operation of the microgrid. Though dq conversion, the transient voltage is converted to U_{dMG} and U_{qMG}, and the corresponding reference values are 311 and 0. The difference is sent to PI regulator for no difference tracking (i.e., $U_{dMG} = 311$ V, $U_{qMG} = 0$). Under this condition, the voltage phase of the microgrid system is equal to θ_{MG}. Therefore the regulation to the actual voltage phase of the microgrid is the regulation to the θ_{MG}.

To realize presynchronization of a microgrid system, It is necessary to carry out θ_{MG} step approaching according to the difference between θ_{MG} and the θ_g, and carry out U_{dMG} step approaching according to the difference between U_{dMG} and grid voltage amplitude U_{dg}. When the amplitude and the phase differences fall within the allowable range, the reference voltage amplitude of the microgrid is assigned to the voltage amplitude at grid side U_{dMG}, and the phase θ_g is delivered to the coordinates conversion process. In this case, the microgrid is running off-grid with the grid voltage amplitude and phase, realizing synchronization of the voltage amplitudes and phases at both sides of the system.

Formula (3.30) to Formula (3.32) are the principles of amplitudes and phases step approaching algorithm.

$$\begin{cases} \theta'_{MG} = \theta_{MG} - step_\theta{}^* \operatorname{sign} \\ \theta_{MG} = \theta'_{MG} \\ \Delta\theta = \theta'_{MG} - \theta_g \end{cases} \tag{3.30}$$

$$\begin{cases} U'_{dMG} = U_{dMG} - step_U{}^* \operatorname{sign} \\ U_{dMG} = U'_{dMG} \\ \Delta U = U'_{dMG} - U_{dg} \end{cases} \tag{3.31}$$

TABLE 3.5 Definitions of variables

Variables	Definitions	Variables	Definitions
θ_{MG}	Voltage phase at MG side before adjustment	$step_\theta$	Phase adjusting step
θ'_{MG}	Phase at MG side after adjustment	$step_U$	Voltage amplitude adjusting step
θ_g	Voltage phase at grid side	$\Delta\theta$	Phase difference
U_{dMG}	Voltage amplitude at MG side		
U'_{dMG}	Real-time voltage amplitude	ΔU	Voltage amplitude difference
U_{dg}	Voltage amplitude at grid side	$sign$	Symbolic variable

$$\text{sign} = \begin{cases} 1, \Delta\theta > 0 \ \ \text{or} \ \ \Delta U > 0 \\ -1, \Delta\theta < 0 \ \ \text{or} \ \ \Delta U < 0 \\ 0, \Delta\theta = 0 \ \ \text{or} \ \ \Delta U = 0 \end{cases} \tag{3.32}$$

Definitions of variables in Formula (3.30) to Formula (3.32) are given in Table 3.5.

Detail adjustment procedures are as follows: when $\Delta\theta > 0$, the voltage of the microgrid system is higher than the voltage at grid side, it has $sign = 1$. In this case, reduce the voltage phase of the microgrid by $step_\theta$; when $\Delta\theta < 0$, the voltage of the microgrid system is lower than the voltage at the grid side, it has $sign = -1$. In this case, increase the voltage phase of the microgrid by $step_\theta$; when $\Delta\theta = 0$ the voltage phases at the two sides are the same, it has $sign = 0$, no approaching adjustment will made. The procedures for voltage amplitude adjustment are similar to those for phase adjustment.

Fig. 3.23 shows the control block diagram of voltage amplitude and phase step approaching adjustment; Fig. 3.24 shows flow chart of the amplitude and phase step approaching algorithm.

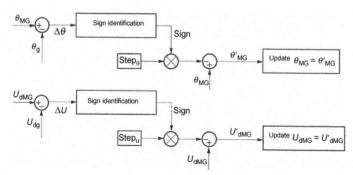

FIG. 3.23   Control block diagram of amplitude and phase step approaching adjustment.

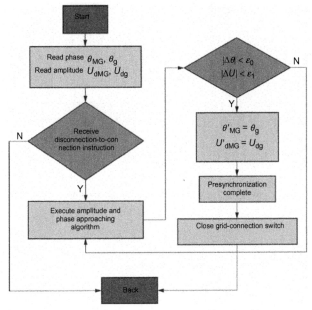

FIG. 3.24   Flow chart of amplitude and phase step approaching algorithm.

TABLE 3.6 Main parameters

Items	Parameter values	Items	Parameter values
DC source voltage (V)	600	Reference MG amplitude (V)	311
Set initial phase difference $\Delta\theta$ (radian)	$\pi/6, \pi$	Phase step: $step_\theta$ (radian)	0.01
Initial status of the grid-connecting switch	Disconnected	Amplitude step: $step_U$ (V)	1

3.4.2   Simulation verification

Establishing a system simulation model under the MATLAB/Simulink simulation platform, where a DC power source is provided at the DC side, and the V/f control mode is adopted for grid-disconnected operation of the microgrid system. Main calculation parameters are set out in Table 3.6.

3.4.2.1   Phase approaching adjustment

Set the reference voltage amplitude U_{dMG} of the microgrid system to the same voltage amplitude as the grid U_{dg} (i.e., $U_{dMG} = U_{dg} = 311\,V$). Besides execute

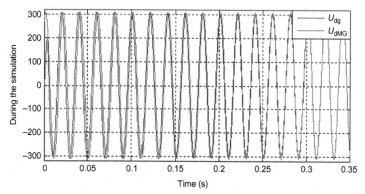

FIG. 3.25 Voltage waveform at two sides of the system based on phase step approaching.

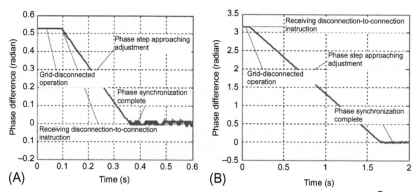

FIG. 3.26 Phase difference diagram based on phase step approaching. (A) Set $\Delta\theta = \dfrac{\pi}{6}$ before adjustment and (B) set $\Delta\theta = \pi$ before adjustment.

the phase step approaching algorithm only, and the simulation results are as shown in Figs. 3.25 and 3.26.

In Fig. 3.25, during the simulation, the phase approaching adjustment is carried out at 0.1 s, and the phase synchronization is realized at about 0.3 s.

As shown in Fig. 3.26, $\Delta\theta$ is set to be $\dfrac{p}{6}$ and π, respectively, comparison on the regulation effects thereof has been performed. According to Fig. 3.26, the larger the $\Delta\theta$ is, the longer adjustment time will take in phase approximation algorithm.

3.4.2.2 Amplitude approaching

Set the grid amplitude as $U_{dg} = 330\,V$, the reference voltage amplitude at the microgrid system side to be 311 V, and the initial voltage phase difference at two sides $\Delta\theta = 0$. Execute the amplitude approaching algorithm, and the

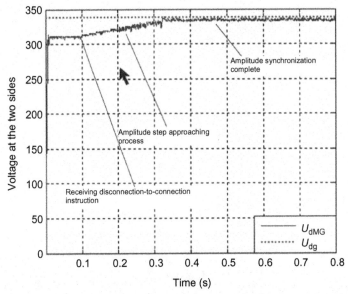

FIG. 3.27   *D*-axis voltage component based on amplitude step approaching adjustment.

simulation results are as shown in Fig. 3.27. In Fig. 3.27, the amplitude approaching adjustment is carried out at 0.1 s, and the amplitude synchronization is realized at about 0.3 s.

3.4.3 Experimental verification

The testing system is as shown in Fig. 3.28, a simulation grid is provided at the grid side, where K1 stands for the grid connecting switch, a PV simulator is provided at the DC side of the PV inverter, and the energy storage converter adopts two-stage topological structure, with the prestage connected to the storage battery. The experimental parameters are as shown in Table 3.7.

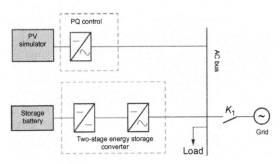

FIG. 3.28   Microgrid test platform.

TABLE 3.7 Main testing parameters

Items	Parameter values	Items	Parameter values
Power of PV inverter	5 kW	Phase modulation step: $step_\theta$	0.02 (radian)
Effective voltage of main power sources	140 V	Amplitude modulation step: $step_U$	4 V
Effective voltage of simulated grid	85 V	Load	10 kW

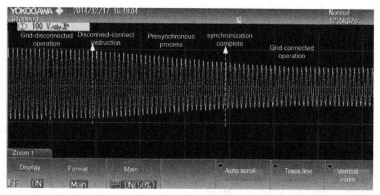

FIG. 3.29 Bus voltage at microgrid side for disconnection-to-connection switching.

The testing results in relation to connection-to-disconnection switching of the energy storage converter are as shown in Fig. 3.29. First, the energy storage converter operates off-grid in the V/f control mode, with K1 disconnected. After the instruction of disconnection-to-connection switching is received, the presynchronization control starts. Start regulating the voltage and phase at the main power source side; and it shows the fluctuation is minor. After synchronization, K1 closes and the energy storage converter is connected to the grid, and the smooth and surge-free disconnection-to-connection switching is realized.

3.5 Self-optimization virtual synchronous generator technology

During grid-connected operation, the traditional PV inverter and energy storage converter operate under P/Q mode, and are not involved in grid frequency and voltage regulation. In the case of low DG connecting capacity and low capacity penetration, the grid will supply stable voltage and frequency, while in the case of large DG connecting capacity and high capacity penetration, too many inertia-free distributed resources may affect the stable operation of the grid. With the synchronous generator control technology, the inverter is endued with the sound internal and mechanical characteristics of the synchronous generator, thus it is able to actively participate in grid regulation, which can effectively improve the stability domain for grid connection and increase the connecting capacity penetration of the distributed resources.

After design and manufacture of a synchronous generator, the rotational inertia and damping of such generator cannot be changed. Therefore, the rotational inertia and the damping of the traditional synchronous generator are fixed. If the distributed resources are designed to have fixed rotational inertia and damping as the traditional generator does, numerous distributed resources with fixed rotational inertia and damping will result in total rotational inertial increase of the system, which may lead to slow dynamic response; Therefore numerous distributed resources with fixed damping will result in damping increase and prolong the transient process of the system.

This book introduces a self-optimization virtual synchronous generator technology. With this technology, the characteristics of the power electronic resources of being flexible and controllable can most advantageous. The generator possesses the rotational inertia and the damping as the synchronous generator possessed, participating in grid regulation, and overcomes the shortcoming of the synchronous generator with fixed rotational inertia and the damping. In case of large disturbance to the voltage frequency, the rotational inertial and the damping will increase, while in case of small disturbance to the voltage frequency, the rotational inertial and the damping will decrease—that is, the rotational inertial and the damping self-adapt to the frequency disturbance (i.e., when the system frequency deviation is large, DG with rotational inertia will cause the rotational inertia increase of the whole grid system and smooth system frequency change, while a damped DG may increase the damping of the whole grid system, so as to shorten the transient process for system frequency change). In microgrid applications, DG with inertia will enhance the effect of rotational inertia of the whole microgrid and significantly improve the operation stability of the microgrid during grid-disconnected operation. This can realize friendly grid connection of distributed resources and "plug-and-play" microgrid connection, and meet the requirements of connecting the distributed resources to the grid with high capacity penetration.

3.5.1   Virtual synchronous generator technology

During grid-connected operation of the traditional power electronic resources, in the P/Q mode, the frequency of the grid is determined by the macrogrid. The virtual synchronous generator (VSG) is used to simulate the traditional synchronous generator during grid connection of the power electronic resources, which is featured by high impedance, large inertia, and self-synchronization. Such

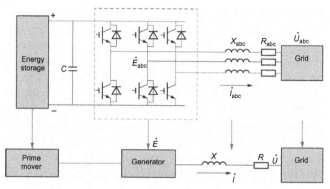

FIG. 3.30   Equivalent relation between converter and synchronous generator.

characteristics are implanted into the control strategy of the distributed resources, so that the external characteristics of the same are equivalent to those of the synchronous generator, which can participate in grid frequency and voltage regulation.

The equivalent relation between the converter and the synchronous generator is as shown in Fig. 3.30. The topological structure of the converter simulates the characteristics of the traditional synchronous generator, and the mechanical equation is as follows:

$$
\begin{cases}
J\dfrac{d\omega}{dt} = \dfrac{P'_T - P_{VSG}}{\omega} - D(\omega - \omega_N) \\
\omega = \dfrac{d\theta}{dt}
\end{cases}
\tag{3.33}
$$

where J stands for the inertia time constant of the virtual generator; P'_T and P'_{VSG} stand for the input virtual mechanical power and the virtual electromagnetic power; ω stands for the angular speed of the rotor of the virtual synchronous generator; ω_N stands for the synchronous angular speed of the grid (rad/s); D stands for damping coefficient (N·m·s/rad); and θ stands for the radian (rad).

The electromagnetic equation of the virtual synchronous generator is as follows:

$$
\dot{E}_{abc} = (R_{abc} + jX_{abc})\dot{I}_{abc} + \dot{U}_{abc}
\tag{3.34}
$$

where $\dot{E}_{abc}$ stands for the output voltage at the inverter bridge side, which is equivalent to the electromotive force of the synchronous generator. $\dot{U}_{abc}$ stands for the voltage at turbine side of the virtual synchronous generator. R_{abc} and X_{abc} stand for the synchronous resistance and the synchronous reactance of the virtual synchronous generator. The virtual synchronous generator possesses the external characteristics of the voltage source, which has two operation modes: the grid-connected mode and the grid-disconnected mode. Therefore, it can maintain the initial state for grid connection during planned islanding or unplanned islanding, which can realize seamless switching from grid connection to grid disconnection.

3.5.2   Self-optimization virtual synchronous generator technology

The self-optimization virtual synchronous generator technology refers to the self-adoption of the rotational inertia J and the damping coefficient D to the frequency fluctuation degree, so as to realize parameter optimization:

$$\begin{cases} J = J_0 + k_J{}^*(|f_N - f|) \\ |f_N - f| = \Delta f \end{cases} \tag{3.35}$$

$$\begin{cases} D = D_0 + k_D{}^*(|f_N - f|) \\ |f_N - f| = \Delta f \end{cases} \tag{3.36}$$

where f_N stands for the rated frequency of the grid; f stands for the actual frequency of the system; k_J stands for the rotational inertia adaptive proportional constant; k_D stands for the damping adaptive proportional constant; J_0 stands for the initial setting of the rotational inertia; and D_0 stands for the initial setting of the damping.

In Formula (3.35), the rotational inertial J is directly in proportion to $|\Delta f|$, which means in the event the frequency fluctuation $|\Delta f|$ of the system is severe, the rotational inertial J will increase, which further drives active power change increasing. In this case, frequency change of the system is getting smooth, which strengthens system frequency support and improves system stability. When $|\Delta f|$ is small, J will decrease. Therefore, J value will not become too high or too low, which may result in too fast or too slow a dynamic response of the system.

In Formula (3.36), the damping D is directly in proportion to $|\Delta f|$, which means that in the event the frequency fluctuation $|\Delta f|$ of the system is large, the damping D will increase, which will shorten the transient process. When the frequency fluctuation $|\Delta f|$ of the system is small, the damping D will decrease, which will prolong the transient process. Therefore, damping D of the system will not become too high or too low, which may affect the transient process of the system.

3.6   Bus dominant hybrid microgrid coordination control technology

The microgrid can serve as a controllable unit of the distribution grid. When the microgrid is disconnected from the distribution grid due to failure, the microgrid can maintain the internal power supply until the failure is eliminated and meet the power quality requirements of the customer. The microgrid can be divided into AC microgrid, DC microgrid, and AC&DC hybrid microgrid according to the bus properties, where the AC&DC hybrid microgrid has both AC bus and DC bus, and can supply directly to the AC loads and to the DC loads. It can exert the advantages of the AC microgrid and the DC microgrid simultaneously. The AC&DC hybrid microgrid is equipped with the advantages of AC microgrid, such as mature technology, high power quality, and sound load compatibility. Besides it also has the advantages of DC microgrid such as low cost of interface device, high energy utilization, and strong robustness. The supply mode of the AC&DC hybrid microgrid facilitates the integration of all kinds of distributed generation, which is an effective approach to solving the problem of grid connection of intensive distributed resources.

3.6.1 Bus dominant division

The structure of the AC&DC hybrid microgrid is as shown in Fig. 3.31. The AC bus of the AC microgrid and the DC bus of the DC microgrid are connected via a two-way power electronic device, to realize the interconnection between the AC microgrid and the DC microgrid. The power electronic device used for such connection is the Hybrid Microgrid Flow Conditioner (HMFC). The HMFC realize voltage and flow control on the DC bus and AC bus of the AC&DC hybrid microgrid via a two-way DC/AC converter. The PCS in the figure is a power conversion system consisting of an energy storage converter.

According to the state of PCC for grid connection, the AC&DC hybrid microgrid can be divided into AC bus dominant type, DC bus dominant type, and auto-dominant type. The "AC bus dominant" hybrid microgrid is as shown in Fig. 3.31A, where the PCC1 is connected to the AC distribution grid via the AC bus and HMFC is used for controlling voltage stability on the DC bus. Compared with the DC bus, the AC bus is in a dominant position and hence it is called a "AC bus dominant" hybrid microgrid. The "DC bus dominant" hybrid microgrid is as shown in Fig. 3.31B where the PCC2 is connected to the DC distribution grid via the DC bus; HMFC is used for controlling voltage stability on the AC bus. Compared with the AC bus, the DC bus is in a dominant position and hence it is called a "DC bus dominant" hybrid microgrid. Considering the coexistence of AC grid and DC grid in the future, the "self-dominant" hybrid microgrid is proposed as shown in Fig. 3.31C.

The PCC is connected to the AC grid via PCC1 and connected to the DC grid via PCC 2; therefore, this type of hybrid microgrid belongs to the "self-dominant" hybrid microgrid.

3.6.2 Dominant control strategy

3.6.2.1 Dominant strategy

"Dominant strategy" is a term in game theory that refers to the optimal option for a player among all the competitive strategy set, no matter how that player's opponents may play, and the opposite strategy is called "inferior strategy."

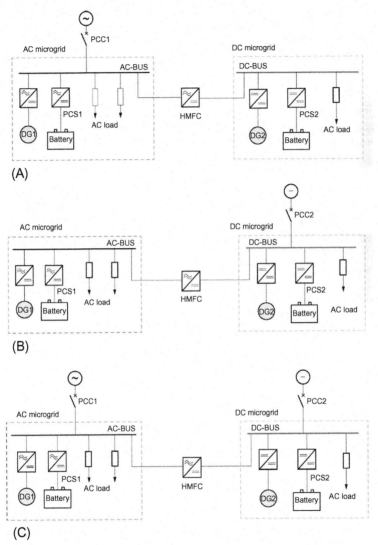

FIG. 3.31 Structure of AC&DC hybrid microgrid. (A) Structure of "AC bus dominant" hybrid microgrid; (B) structure of "DC bus dominant" hybrid microgrid; and (C) structure of "self-dominant" hybrid microgrid.

The AC&DC hybrid microgrid has two buses with different types of power sources. Bus control is a key factor to ensure stable operation of the AC&DC hybrid microgrid. Game theory is introduced into the mode control of the AC&DC hybrid microgrid. During operation, the AC microgrid and the DC microgrid play a game. The bus that supports the stable operation of the whole hybrid microgrid is the dominant party, called the dominant bus, while the other bus is in the subordinate position, and called the subordinate bus. The dominant bus controls the subordinate bus via HMFC, to realize stable operation of the AC&DC hybrid microgrid.

3.6.2.2 Operation mode

For the AC&DC hybrid microgrid, there are two operation modes: grid-connected mode and grid-disconnected mode. In the grid-connected mode, the "AC bus dominant" hybrid microgrid and the "DC bus dominant" hybrid microgrid are connected to the distribution grid via a unique PCC, similar to the grid-connected operation of a single microgrid. For the "self-dominant" hybrid microgrid, there are two PCCs for grid connection. HMFC serves as a flexible switch, which is used to connect the DC microgrid and the AC microgrid and control the power flow between the microgrids. When PCC is disconnected, the case is equivalent to grid-connected operation of an "AC bus dominant" or a "DC bus dominant" hybrid microgrid. In grid-disconnected mode, when disconnecting the PCC, all three types of AC&DC hybrid are equivalent to a grid-disconnected DC microgrid and a grid-disconnected AC microgrid connected via the HMFC, as shown in Fig. 3.32.

When the AC microgrid and the DC microgrid are operating separately, it is equivalent to an independent AC microgrid and an independent DC microgrid. For both the AC microgrid and the DC microgrid, the HMFC can be simplified to a DG unit. The flow control is subject to the scheduling of the energy management system.

3.6.2.3 Dominant criteria

In the grid-connected operation mode, the distribution grid to be connected will serve as the dominant criteria. In the grid-disconnected operation mode, the AC microgrid and the DC microgrid compete with each other and also serve as a backup for each other, and the optimal operation mode is realized via gaming competition. Shifting between the operation modes is a random and multiattribute decision-making process. The dominant relation is identified on basis of the dominant criteria, so as to determine the dominant strategy.

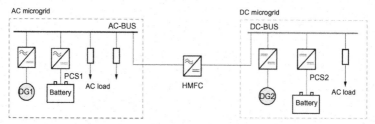

FIG. 3.32 Structure of AC&DC hybrid microgrid under grid-disconnected operation.

In the grid-disconnected mode, the AC&DC hybrid microgrid is not supported by the distribution grid with infinite capacity. In this case, the energy storage system will be provided for transient power balance between DG and loads, so as to maintain the stable operation of the microgrid. Limited by the randomness of DG, in grid-disconnected operation, only the energy storage can be reliably scheduled by the microgrid. Like the fuel indicator of a car, when a car is driving along a road in a deserted area where the location of a gas station is unknown, the fuel indicator is a key base for determining whether such a trek is possible. Similarly, for the grid-disconnected operation of the AC&DC hybrid microgrid, energy storage status will serve as the dominant criteria.

For the AC&DC hybrid microgrid, SOC of the energy storage indicates the power state. SOC $=0$ indicates the battery is out of power, and SOC $=1$ indicates the battery is fully charged. The state of charge of the ES of the AC microgrid is indicated as SOC_{AC}. In the grid-disconnected mode, the allowable upper limit is SOC_{AC_H} and the lower limit is SOC_{AC_L}, while the state of charge of the ES of the DC microgrid is indicated as SOC_{DC}. In the grid-disconnected mode, the allowable upper limit is SOC_{DC_H} and the lower limit is SOC_{DC_L}. In the grid-disconnected mode, the dominant criteria for operation mode switching will meet the requirements of Formula (3.37).

$$\text{Dom}(t) = \begin{cases} M_{AC}(SOC_{DC} < SOC_{DC\_L}) \& (SOC_{AC} > SOC_{AC\_L}) \\ M_{DC}(SOC_{AC} < SOC_{AC\_L}) \& (SOC_{DC} > SOC_{DC\_L}) \\ 0 \end{cases} \qquad (3.37)$$

In Formula (3.37), M_{AC} stands for AC bus dominant, M_{DC} stands for DC bus dominant. For grid-disconnected operation of a hybrid microgrid, when SOC_{DC} is lower than SOC_{DC_L} and SOC_{AC} is higher than SOC_{AC_L}, the dominant strategy is M_{AC}; When SOC_{AC} is lower than SOC_{AC_L} and SOC_{DC} is higher than SOC_{DC_L}, the dominant strategy is M_{DC}, otherwise there will be no dominant strategy, and the operation mode of the hybrid microgrid will maintain unchanged.

3.6.3   Grid-connected operation

3.6.3.1   AC bus dominant type

As shown in Fig. 3.31A, the hybrid microgrid is connected to the AC grid via PCC1. In the grid-connected mode, the AC grid provides AC bus voltage and frequency supports, and HMFC operates in the DC voltage source mode (the V control mode) for controlling voltage stability on the DC bus. The power flow within the AC&DC microgrid will be distributed freely based on the load demands and distributed generation output condition; DC loads are supplied by DG and ES on the DC microgrid, surplus energy will be transmitted to the AC microgrid via HMFC, while in case of power insufficiency, power will be supplied by AC microgrid via HMFC.

3.6.3.2 DC bus dominant type

As shown in Fig. 3.31B, the hybrid microgrid is connected to the DC grid via PCC2. In the grid-connected mode, the DC grid provides DC bus voltage supports, and HMFC operates in the AC voltage source mode (the V/f control mode) for controlling voltage and frequency on the AC bus. In the power flow within the AC&DC microgrid, power will be distributed freely based on the load demands and distributed generation output condition. The AC loads are supplied DG and ES on the AC microgrid, surplus energy will be transmitted to the DC microgrid via HMFC, while in case of power insufficiency, power will be supplied by DC microgrid via HMFC.

3.6.3.3 Auto-dominant type

As shown in Fig. 3.31C, the "self-dominant" hybrid microgrid has two PCCs, which are connected to the AC grid and the DC grid via PCC1 and PCC2, respectively. In the grid-connected mode, HMFC serves as a flexible switch, which is used to connect the DC microgrid and the AC microgrid. HMFC operates in the power flow control mode (the P/Q control mode), and power flow is subject to the scheduling of the energy management system. When PCC1 is disconnected, it is equivalent to the grid connected operation mode of "DC bus dominant" hybrid microgrid, and HMFC will switch to the V/f control mode for controlling the voltage and frequency on the AC bus; When the PPC2 is disconnected, it is equivalent to the grid-connected operation mode of "AC bus dominant" hybrid microgrid, and HMFC will switch to the V control mode for controlling the voltage stability on the DC bus.

3.6.4 Grid-disconnected operation

3.6.4.1 AC bus dominant type

When the hybrid microgrid is disconnected from the grid, PCS1 works in the V/f control mode, to control the voltage and frequency of the AC bus, and HMFC remains working in the V control mode, to maintain the voltage stability on the DC bus. PCS2 remains working in the current source mode (the I control mode), to provide power supports for the DC microgrid. Power should be distributed freely between the AC microgrid and DC microgrid; according to the load demands and the DG outputs, the working mode satisfies Formula (3.38).

$$M_{\text{AC}} = \begin{cases} \text{PCS1} : v/f\,\text{control mode} \\ \text{HMFC} : V\,\text{control mode} \\ \text{PCS2} : I\,\text{control mode} \end{cases} \tag{3.38}$$

Formula (3.38) reflects the classification basis for "AC bus dominant; M_{AC} is the main mode during grid-disconnected operation of "AC bus dominant" hybrid microgrid."

In case of PCS1 failure, or $SOC_{AC} < SOC_{AC\_L}$, and PCS2 is working in normal conditions and $SOC_{DC} > SOC_{Dc\_L}$, the dominant strategy will swithc to M_{DC}, PCS1 will stop working or shift to P/Q control mode, HMFC will switch to V/f control mode, taking over the control over the AC bus, and PCS2 will switch to V control mode, taking over the control over the DC bus. This operation mode is supported by DC microgrid, which is a standby mode for grid-disconnected operation of the "AC bus dominant" hybrid microgrid.

In case of PCS2 failure or $SOC_{DC} < SOC_{Dc\_L}$, when PCS1 is working in normal conditions and $SOC_{AC} > SOC_{AC\_L}$, the hybrid microgrid will carry out main-standby mode recovery—i.e., the operating mode shifts from M_{DC} to M_{AC}; PCS1 shifts to V/f control mode, restoring control over the AC bus and maintaining the voltage and frequency stability on the AC bus; HMFC shifts to V control mode, restoring control over the DC bus; and PCS2 shifts to I control mode, completing main-standby mode recovery for grid-disconnected operation of the "AC bus dominant" hybrid microgrid. Procedures for the main-standby mode shifting and the main-standby mode recovery of the "AC bus dominant" hybrid microgrid are as shown in Fig. 3.33.

3.6.4.2 DC bus dominant type

When the hybrid microgrid is disconnected from the grid, PCS2 works in the V control mode, taking over control of the AC bus. HMFC remains working in the V/f control mode, to control the voltage and frequency of the AC bus, and PCS1

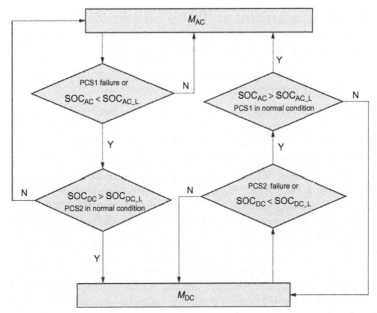

FIG. 3.33 Grid-disconnected operating mode shifting of "AC bus dominant" hybrid microgrid.

works in the P/Q control mode, to provide power supports to the AC microgrid. Power should be distributed freely between the AC microgrid and DC microgrid according to the load demands and the DG outputs, and the working mode satisfies Formula (3.39).

$$M_{DC} = \begin{cases} \text{PCS1} : \text{P/Qcontrol mode} \\ \text{HMFC} : \text{V/fcontrol mode} \\ \text{PCS2} : \text{Vcontrol mode} \end{cases} \quad (3.39)$$

Formula (3.39) reflects the classification basis for DC bus dominant, M_{BC} is the main mode during grid-disconnected operation of the "DC bus dominant" hybrid microgrid.

In case of PCS2 failure, or when $SOC_{DC} < SOC_{DC\_L}$ and PCS1 is working in normal condition and $SOC_{AC} > SOC_{AC\_L}$, the dominant strategy will shift to M_{AC}, PCS2 will stop working or shift to the I control mode, HMFC will shift to V control mode, to take over the control over the DC bus, and PCS1 will shift to V/f control mode, to take over the control over the AC bus. This operation mode is supported by AC microgrid, which is a standby mode for grid-disconnected operation of the "DC bus dominant" hybrid microgrid.

In case of PCS1 failure or $SOC_{AC} < SOC_{AC\_L}$, and PCS2 is working under normal conditions and $SOC_{Dc} > SOC_{Dc\_L}$, the hybrid microgrid will carry out main-standby mode recovery—i.e., the operating mode shifts from M_{AC} to M_{DC}, PCS2 shifts to V control mode, restoring control over the DC bus, HMFC shifts to V/f control mode, restoring control over the AC bus and frequency, and PCS1 shifts to P/Q control mode, completing main-standby mode recovery for grid-disconnected operation of the "DC bus dominant" hybrid microgrid. Procedures for main-standby shifting and main-standby recovery of the"DC bus dominant" hybrid microgrid are as shown in Fig. 3.34.

3.6.4.3 Auto-dominant type

When the two PCCs are disconnected in sequence, the operating mode will perform auto-dominant according to the disconnecting sequence, which reflects the classification basis for the "auto-dominant type" If PPC1 disconnected first, HMFC will shift to the V/f control mode, to maintain the voltage and frequency stability of the AC bus. After PPC2 disconnects, the procedures are the same as those for shifting from grid connection to grid disconnection of the "DC bus dominant" hybrid microgrid. If PCC2 disconnected first, HMFC will shift to the V control mode, to control voltage stability on the DC bus. After PPC1 disconnects, the procedures are the same as those for shifting from grid connection to grid disconnection of the "AC bus dominant" hybrid microgrid.

If two PCCs are disconnected simultaneously (i.e., the AC microgrid and the DC microgrid are shifted to grid-disconnected operation mode simultaneously), PCS1 will shift to the V/f control mode for controlling voltage and frequency stability on the AC bus, PCS2 will shift to the V control mode, to control the

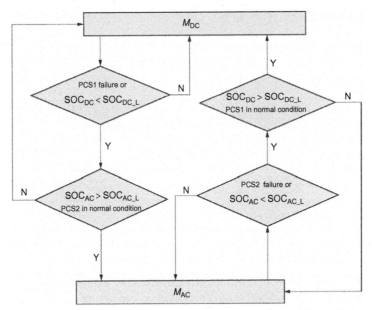

FIG. 3.34 Grid-disconnected operating mode shifting of "Dc bus dominant" hybrid microgrid.

voltage stability on the DC bus, and HMFC will shift to the standby mode and subjected to the scheduling of the energy management system. After stable operation, the operating mode of HMFC is subject to the "auto-dominant" control based on dominant criteria, or is subject to the scheduling of the energy management system according to the energy optimizing operation control strategy, regulating the power flow of HMFC.

3.6.5 Grid connection recovery

The procedures for shifting from grid-disconnected operation to grid-connected operation of the "AC bus dominant" and "DC bus dominant" hybrid microgrids are the same as those for grid-connection of single microgrid, and the operating mode of HMFC remains unchanged; When reconnecting to the grid from the standby mode, for the "AC bus dominant" hybrid microgrid, HMFC shifts from the V/f control mode to the V control mode, to control the voltage on the DC bus, and PCS2 shifts to the I control mode, grid re-connection complete. For the "DC bus dominant" hybrid microgrid, HMFC shifts from the V control mode to the V/f control mode, controlling the voltage and frequency on the AC bus, and PCS2 shifts to the P/Q control mode, grid re-connection complete.

The two PCCs of the "auto-dominant" hybrid microgrid should be reconnected to the grid one by one. When the second PPC is reconnected to the grid, HMFC shifts from the V/f control mode or the V control mode to the P/Q control mode as a flexible contacting switch.

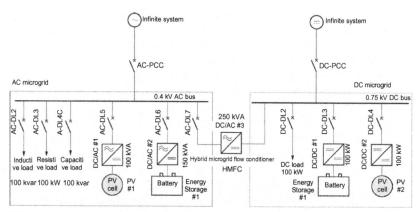

FIG. 3.35   Topological structure of AC&DC hybrid microgrid.

3.6.6   Simulation verification

Fig. 3.35 shows the topological structure of AC&DC hybrid microgrid established for simulation experiment. This structure consists of an AC microgrid system, a DC microgrid system and a HMFC for power conversion between the AC bus and the DC bus. The AC microgrid and the DC microgrid are connected to the external grid via the PCC (AC-PCC and DC-PCC, corresponding to the PCC1 and PCC2 in Fig. 3.31). The AC&DC hybrid microgrid can operate in the AC microgrid separate operating mode, the DC microgrid separate operating mode and AC&DC microgrids interconnecting operation mode. In the simulation experiment, the voltage of AC microgrid bus is 400 V; the inductive loads, the resistive load, and the capacitive load connected to the grid are 100 kvar, 100 kW, and 100 kvar, respectively; the capacity of PV #1 DC/AC is 100 kVA; the capacity of Energy Storage #1 (DC/AC) is 150 kVA; the voltage of the DC microgrid bus is 750 V; DC loads connected are 100 kW; the capacity of Energy Storage #2 (DC/ DC) is 100 kW; and the capacity of PV #2 (DC/DC) is 100 kV. The capacity of Flow Controller 3 (DC/AC) between the AC microgrid and the DC microgrid is 250 kVA.

3.6.6.1   Grid-connected loads switching experiment

For the grid-connected operation of the "auto-dominant" hybrid microgrid, both the DC side and AC side are supported by macrogrid, and no further description is given herein. This book takes the "AC bus dominant" and "DC bus dominant" hybrid microgrids as an example for illustrations. Table 3.8 shows the full capacity load switching experiment with maximum load change rate during the grid-connected operation of the AC&DC hybrid microgrid.

Procedures for the simulation of grid-connection of the AC&DC hybrid microgrid are as follows:

TABLE 3.8 Grid-connected loads switching experiment

Items	Content
"AC bus dominant" hybrid microgrid DC load switching experiment (Energy Storage #2 and PV #2 are out of service)	100 kW → 0 kW
	0 kW → 100 kW
"DC bus dominant" hybrid microgrid AC load switching experiment (Energy Storage #1 and PV #1 are out of service)	100 kW → 0 kW
	0 kW → 100 kW

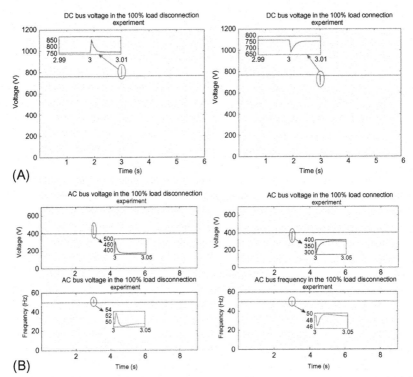

FIG. 3.36 DC bus voltage waveform and AC bus voltage and frequency waveforms in AC&DC hybrid microgrid load switching experiment.

(1) "AC bus dominant" hybrid microgrid: simulation is performed when the DC-PCC is disconnected, The HMFC controls the voltage of DC microgrid bus, and DC load switching is performed at 3 s. The whole simulation process lasts for 6 s.

Fig. 3.36A shows the voltage waveform of the DC bus during the simulation process from the locally enlarged view in Fig. 3.36A. During the

DC load switching process at 3 s, DC bus voltage decrease (increase) temporarily, with the maximum deviation at about 85 V and overshoot of 11%. However the whole regulation is complete within 3 ms. According to the two figures, in case of voltage fluctuation, the V control mode of HMFC can effectively maintain the voltage stability on the DC bus.

(2) "DC bus dominant" hybrid microgrid: simulation is performed when the AC-PCC is disconnected. The HMFC controls the voltage and frequency of the AC microgrid bus, and AC load switching is performed at 3 s. The whole simulation process lasts for 6 s.

Fig. 3.36B shows the voltage and frequency waveform of the AC bus during the simulation process. According to the two voltage waveforms in Fig. 3.36B, during load switching at 3 s, the maximum voltage deviation on the AC bus is 100 V with an overshoot of 25%, and the maximum frequency deviation is 3.7 Hz (3Hz) with an overshoot of (7.4%) 6%, and the whole regulation lasts for about 10 ms. Results show that in the "DC bus dominant" hybrid power grid, the V/f control mode of HMFC can effectively and rapidly stabilize the voltage and frequency on the AC bus in case of voltage and frequency fluctuation on AC microgrid, and ensure the stable operation of the system.

3.6.6.2 Grid connection-disconnection switching experiment

The grid connection-disconnection switching experiment of the "auto-dominant" hybrid microgrid covers the grid connection-disconnection switching of "AC bus dominant" and "DC bus dominant" hybrid microgrids. Therefore, no separate descriptions are provided herein.

During the disconnection to connection process of the AC&DC hybrid microgrid, according to the closing sequence of the AC-PCC and DC-PCC, there will be three operating conditions. In this process, the hybrid microgrid is supported by the external grid. No obvious oscillation will occur. Therefore no more specific descriptions are provided herein. Here, we only discuss the condition when AC-PCC at AC side and DC-PCC at DC side are closed simultaneously. Experimental operating conditions are set out in Table 3.9.

The simulation procedures for grid connection-disconnection of "auto-dominant" hybrid microgrids are as follows:

(1) At first, the microgrid is operated in the grid-connected mode. Disconnect AC-PCC at AC side at 3 s, disconnect DC-PCC at DC side at 6 s, and connect the two PPCs at AC side and DC sides simultaneously at 9 s. The simulation process lasts for 15 s, and the simulation waveform is as shown in Fig. 3.37.

According to Fig. 3.37A, when AC-PCC is disconnected at 3 s and DC-PPC is disconnected at 6 s, the maximum voltage deviation on the AC bus is 85 V with an overshoot of 21.25%, the maximum frequency deviation is 5 Hz with an overshoot of 10%, and both regulations are completed within 30 ms. The maximum voltage deviation on the DC bus is 30 V with an overshoot of 4%,

TABLE 3.9 Initial operating conditions for the hybrid microgrid in grid connection and disconnection experiment.

PV #1	AC load	Energy storage #1 "+" for discharging; "−" for charging	AC&DC hybrid microgrid flow conditioner	PV #2	DC load	Energy Storage #2 "+" discharging; "−" charging
100 kW	100 kVA (power factor: 0.6)	−100 kW	Supply active power to AC grid	100 kW	100 kW	−100 kW

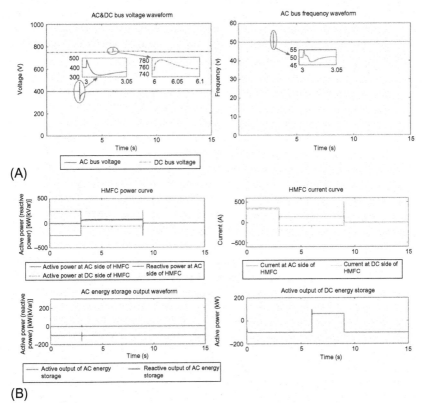

(A)

(B)

FIG. 3.37 Grid connection-disconnection switching experiment of ac&dc hybrid microgrid— Scheme 1.

which hardly impacts the microgrid. When closing AC-PCC and DC-PCC at 9 s simultaneously, the hybrid microgrid is reconnected to the grid. HMFC and Energy Storage #2 shift to the P/Q (and P) control mode from the V/f control mode and the V control mode simultaneously. As supported by the macrogrid, the transient process is smooth.

(2) At first, the microgrid is operated in the grid-connected mode, disconnect DC-PCC at DC side at 3 s, disconnect AC-PCC at AC side at 6 s, and connect the two PPCs at AC side and DC side simultaneously at 9 s. The simulation process lasts for 15 s, and the simulation waveform is as shown in Fig. 3.38.

According to Fig. 3.38A, when disconnecting DC-PCC at 3 s and AC-PCC at 6 s, the maximum voltage deviation on DC bus is 12 V with an overshoot of 1.57%; the impact imposed on waveform can be ignored. The maximum voltage deviation on the AC bus is 125 V with an overshoot of 31.25%, and the maximum frequency deviation is 4.5 Hz with an overshoot of 9%. The regulation and transient process completes within 50 ms. When AC-PCC and DC-PCC are closed at 9 s simultaneously, the hybrid microgrid is reconnected to the grid. HMFC and Energy Storage #1 shift to the P/Q control mode from the V control mode and the V/f control mode simultaneously. As supported by the macrogrid at AC and DC sides, the transient characteristics is ideal.

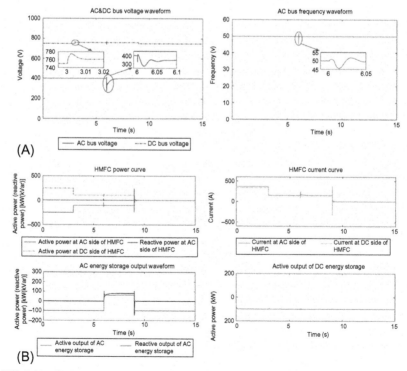

FIG. 3.38 Grid connection-disconnection switching experiment of hybrid microgrid—Scheme 2.

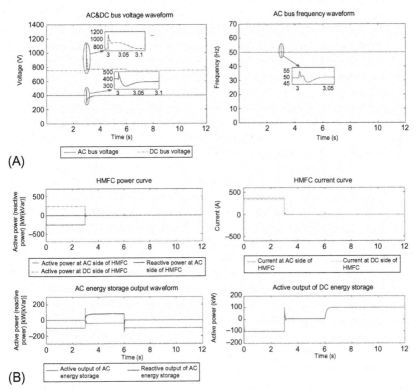

FIG. 3.39 Grid connection-disconnection switching experiment of hybrid microgrid—Scheme 3.

(3) At first, the microgrid operates in the grid-connected mode; disconnect the PCCs at DC side and AC side at 3 s simultaneously, and connect the PPCs at AC side and DC side simultaneously at 6 s. The simulation process last for 12 s, and the simulation waveform is as shown in Fig. 3.39.

As shown in Fig. 3.39A, when AC-PCC and DC-PCC are disconnected at 3 s, HMFC maintains in the P/Q control mode, and Energy Storage #1 shifts to the V/f control mode. The maximum voltage deviation on AC bus is 105 V with an overshoot of 26.25%, and the maximum frequency deviation is 5 Hz with an overshoot of 10%. Both regulations complete within 20 ms.

Energy Storage #2 shifts from the P control mode to the V control mode for voltage stabilization on DC bus. The maximum deviation is 300 V with an overshoot of 39%, and the regulation complete within 80 ms. When AC-PCC and DC-PCC are closed at 6 s simultaneously, the hybrid microgrid is reconnected to the grid. Energy Storage #1 and Energy Storage #2 shift to the P/Q (and P) control mode from the V/f control mode and the V control mode respectively, and transient characteristics of the same are the same as those of the previous two experiments.

Making a general survey on the power and current at both sides of HMFC and the waveforms of Energy Storages 1 and 2 in Fig. 3.37B, Fig. 3.38B, and Fig. 3.39B, we can see that during the connection-disconnection switching process of the AC&DC hybrid microgrid, the power changes within the system conform to the coordination control principle for the switching process, and after a short-time regulation on the power and current waveforms, the system can rapidly resume to the steady state for operation. This verifies the correctness of system modeling and coordination control strategy, and embodies the sound stability and fast dynamic response of the AC&DC hybrid microgrid system.

3.6.6.3 Grid-disconnection mode switching experiment

During the connection to disconnection switching processes of the AC&DC hybrid microgrid, different AC-PCC and DC-PCC disconnecting sequences will result in different working conditions for grid-disconnected operation. To verify the mutual supports within the AC&DC hybrid microgrid, simulation on switching between the grid-disconnected mode of "AC bus dominant" hybrid microgrid and the grid-disconnected mode of "DC bus dominant" hybrid microgrid has been performed, and the initial operating conditions for the experiment are set out in Table 3.10.

(1) Switching from grid-disconnected operation of "AC bus dominant" hybrid microgrid to grid-disconnected operation of "DC bus dominant" hybrid microgrid.

The experiment contents are as shown in Experiment 1 in Table 3.10. Simulation process description: as shown in Fig. 3.35, the simulation is performed under the working conditions when the DC-PCC is disconnected. The HMFC

TABLE 3.10 Initial operating conditions for the hybrid microgrid in the grid-disconnection switching experiment

	Initial state of AC microgrid			AC&DC hybrid microgrid Flow conditioner HMFC	Initial state of DC microgrid		
No.	PV #1	AC load	Energy Storage #1 "+" discharge "−" charge		PV #2	DC load	Energy Storage #2 "+" discharge "−" charge
1	50 kW	100 kW	0	Steady DC bus voltage	100 kW	100 kW	+ 50 kW
2	100 kW	100 kW	+ 50 kW	Steady AC bus voltage and frequency	50 kW	100 kW	0

works in the V control mode to stabilize voltage on the DC bus; AC-PCC is disconnected at 3 s, Energy Storage #1 shifts from the P/Q control mode to the V/f control mode, to provide voltage and frequency supports to the AC microgrid. At 6 s, HMFC shifts from the V control mode to the V/f control mode, to take the place of Energy Storage #1 and take over the control over the voltage and frequency on the AC bus. Energy Storage #1 shifts from the v/f control mode to the P/Q control mode, and Energy Storage #2 shifts from the P control mode to the V control mode, to provide voltage supports to AC bus. The simulation process lasts for 9 s. The results are as shown in Fig. 3.40.

According to Fig. 3.40A, when the hybrid microgrid disconnects from the grid at 3 s, the maximum voltage deviation on AC bus is 35 V, with an overshoot of 8.75%, and the maximum frequency deviation on AC bus is 1.5 Hz, with an overshoot of 3%, and both regulations complete within 50 ms. The maximum voltage deviation on the DC bus is 13 V, with an overshoot of 1.7%. During

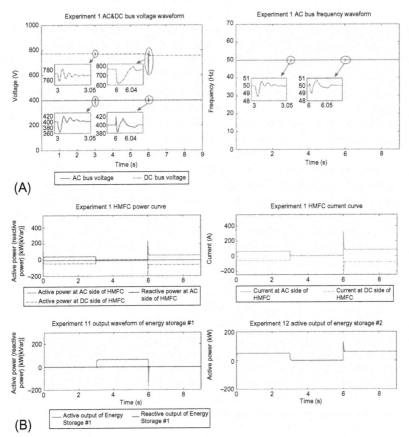

FIG. 3.40 Simulation waveform of grid-disconnected mode switching of the hybrid microgrid in Experiment 1.

the transient process, the transient characteristics are ideal. The hybrid microgrid shifts the operating mode at 6 s, and this process imposes relatively large influence on the DC microgrid. The maximum voltage deviation on DC bus is 141.8 V, with an overshoot of 18.76%, and the regulation duration is 40 ms. The maximum voltage deviation at AC side is 20 V, with an overshoot of 5%. The maximum frequency deviation is 1 Hz, with an overshoot of 2%, and both regulations complete within 60 ms.

(2) Switching from grid-disconnected operation of "DC bus dominant" hybrid microgrid to grid-disconnected operation of "AC bus dominant" hybrid microgrid.

The experiment contents are as shown in Experiment 2 in Table 3.10. Here are some simulation process descriptions: as shown in Fig. 3.35, the simulation is performed under the working conditions when the AC-PCC is disconnected. HMFC works in the V/f control mode, to stabilize voltage and frequency on the AC bus. DC-PCC is disconnected at 3 s. Energy Storage #2 shifts from the P control mode to the V control mode, to provide voltage supports to the DC microgrid. Operating mode shifts at 6 s. HMFC shifts from the V/f control mode to the V control mode, taking the place of Energy Storage #2 and stabilizing the voltage on the DC bus, Energy Storage #2 shifts from the V control mode to the P control mode, and Energy Storage #1 shifts from the P/Q control mode to the V/f control mode, to provide voltage and frequency supports to AC bus. The simulation process lasts for 9 s, and the results are as shown in Fig. 3.41.

As can be seen from Fig. 3.41A, the "DC bus dominant" hybrid microgrid disconnects from the grid at 3 s.

During the process, the maximum voltage deviation on DC bus is 168.8 V, with an overshoot of 21.9%, and the regulation duration is 10 ms. The maximum voltage deviation at AC side is 2.2 V, with an overshoot of 5.5%, the maximum frequency deviation is 1.23 Hz, with an overshoot of 2.46%, and both regulations complete within 20 ms. The hybrid microgrid shifts the operating mode at 6 s. The maximum voltage deviation on DC bus is 237.4 V, with an overshoot of 30.8%, and the regulation duration is 50 ms. The maximum voltage deviation at AC side is 97.4 V, with an overshoot of 24.35%, the maximum frequency deviation is 6.8 Hz, with an overshoot of 13.6%, and both regulations complete within 80 ms.

In addition, power and current of HMFC and the power waveform of Energy Storage #1 and 2 in Fig. 3.40B and Fig. 3.41B show sound transient/steady state characteristics.

3.6.7 Experimental verification

In the experimental physical model, the AC bus voltage is 380 V, the DC bus voltage is 750 V, the power of the energy storage system of the AC microgrid

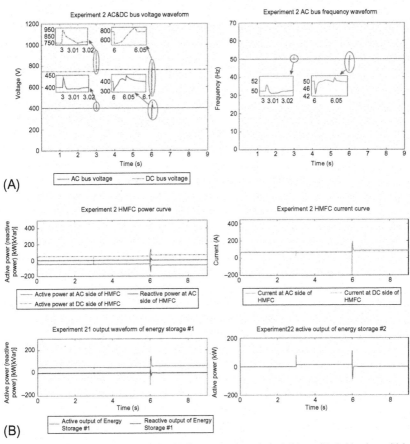

FIG. 3.41 Simulation waveform of grid-disconnected mode switching of hybrid microgrid in Experiment 2.

is 50 kW, the power of the energy storage system of the DC microgrid is 20 kW, and the rated power of the coordinate flow controller of AC&DC microgrid is 50 kW. The main parameters of the experimental system are set out in Table 3.11. Fig. 3.42 is the wiring of the physical model experimental system.

3.6.7.1 AC bus dominance experiment

HMFC works in the V control mode, to stabilize the voltage on DC bus. Regulate the charging and discharging conditions of PCS2 to verify the DC bus voltage control effect of the HMFC. Fig. 3.43 shows the experimental waveform. In the figure, Channel 1 is the voltage at AC side of HMFC, Channel 2 is the voltage at DC side of HMFC, Channel 3 is the current at AC side of HMFC, and Channel 4 is the current at DC side of HMFC. These channel identification are applicable to all experimental waveform diagrams.

TABLE 3.11 Main parameters of AC&DC hybrid microgrid experimental system

Voltage range AC side	400VAC ± 10%	Voltage range DC side	640~760VDC
Maximum current	85A	Maximum voltage	850VDC
Current waveform distortion	Less than 3% (rated power)	Maximum current	96A

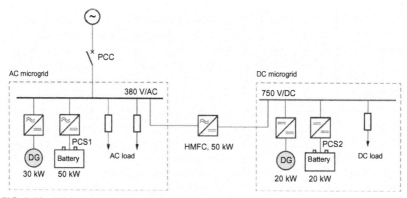

FIG. 3.42 Wiring of AC&DC hybrid microgrid physical model experimental system.

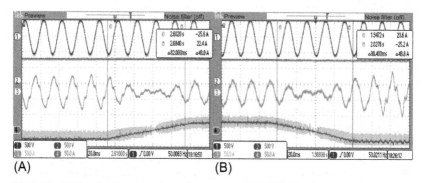

(A)   **(B)**

FIG. 3.43 AC bus dominant experimental waveform. (A) 20 kW Charging to 20 kW discharging and (B) 20 kW discharging to 20 kW charging.

Fig. 3.43A shows the experimental waveform from the experiment of charging by 20 kW to discharging by 20 kW of PCS2. Firstly PCS2 charges the energy storage (20 kW) and the PCS2 shifts to discharging status after 82 ms. The current at DC side of HMFC indicated by Channel 4 changes from −25.6A to 25.4A, and the voltage on DC bus indicated by Channel 2 remains unchanged at 750 V.

Fig. 3.43B shows the experimental waveform from the experiment of discharging by 20 kW to the charging state of PCS2, the current at DC side of HMFC indicated by Channel 4 changes from 25.6A to −25.2A. During the shifting process. The voltage on DC bus indicated by Channel 2 remains unchanged at 750 V, the experimental results verify the function of DC bus voltage support of HMFC in the V control mode.

3.6.7.2   DC bus dominance experiment

HMFC works in the V/f control mode, to control the voltage and frequency of AC bus, and PCS2 works in the V control mode, to control the voltage on DC bus. Switching the loads connected to the AC microgrid to verify the voltage and AC bus frequency control performance of HMFC, and the experimental waveform is as shown in Fig. 3.44.

In Fig. 3.44, when a 20 kW load is connected to the AC side, the DC microgrid supplies 20 kW power to the AC microgrid via the HMFC, and the HMFC is working in the V/f control mode, which can support voltage and frequency on the AC bus. The experimental results verify the function of AC bus voltage and frequency control of HMFC when working in the V/f control mode.

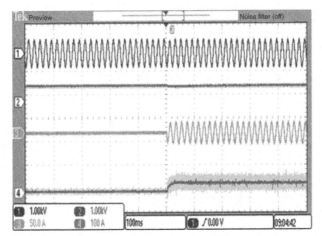

FIG. 3.44 Verification of experimental waveform from DC bus dominant operating mode experiment.

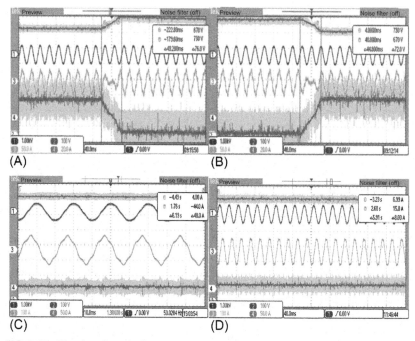

FIG. 3.45   Waveform from the flow controller active/reactive flow control experiment. (A) 20 kW active output to 20 kW active absorption; (B) 20 kW active absorption to 20 kW active output; (C) 50 kvar capacitive reactive output waveform; and (D) 50 kvar inductive reactive output waveform.

3.6.7.3   HMFC active/reactive flow experiment

When the AC microgrid and the DC microgrid are operating interdependently, HMFC outputs power to the AC microgrid or DC microgrid by following the scheduling instruction. The active/reactive control effect of HMFC and the dynamic response time thereof are provided in Fig. 3.45.

In Fig. 3.45A, at 43.2 ms, the AC side of HMFC shifts from the state of outputting 20 kW active power to inputting 20 kW active power. In Fig. 3.45B, at 44.8 ms, the state of HMFC inputting 20 kW active power to AC side shifts to outputting 20 kW active power. Fig. 3.45C shows the waveform of HMFC outputting 50 kvar capacitive reactive power to the AC side; Fig. 3.45D shows the waveform of HMFC outputting 50 kvar inductive reactive power to the AC side.

3.7   Communication line–free interconnecting microgrid control technology

There are mainly three modes for microgrid control, including the master-slave control, the peer-to-peer control, and the integrated hierarchical control.

(1) Master-slave control: during grid-disconnected operation of the microgrid, the main power source (energy storage device) should switch from the P/Q control mode to the V/f control mode. During the grid-connected operation, it should switch from the V/f control mode to the P/Q control mode. For a microgrid adopting the master-slave control, a "gap" exists during the switching process in case of islanding. Although the fast power electronic switch can narrow the "gap," "seamless" switching is impossible. Besides, during grid-disconnected operation, the energy storage device cannot support the large loads for a long time. Even with small loads, it cannot stay in the charging state for a long time. In addition, it relies on the integrated hierarchical control to realize energy balance.

(2) Peer-to-peer control: each distributed resources, according to the voltage and frequency at PCC, adopts droop control to participate in voltage and frequency regulation during grid-disconnected operation of the microgrid. The droop control does not rely on the communication, but how to maintain voltage and frequency stability during grid-disconnected operation remains a problem to be solved. That is the reason why the microgrid with peer-to-peer control is not applied in engineering practice.

(3) Integrated hierarchical control: divide the microgrid into three layers, including the energy management layer, the coordinate control layer, and the local control layer. The MGCC, relying on the coordinate control layer, carries out centralized management over each distributed resource, energy storage device, and load, so as to realize off-grid energy balance of the microgrid. This method is a mature technical control mode that is widely applied to the microgrid with practical commercial applications. The hierarchical control has disadvantages of relying on communication, with complicated structure, low technical index, "gap" switching, unplanned islanding overvoltage, and connecting surge.

3.7.1   Main problems of hierarchical control

The structure of microgrid hierarchical control is as shown in Fig. 3.46, which consists of three control layers, including the energy management layer, the coordinate control layer, and the local control layer. The energy management layer is responsible for the energy management over multiple microgrids within the distribution grid. In the coordinate control layer, MGCC performs centralized management over the distributed resources, energy storage devices and loads. It responds to the scheduling and management of the energy management layer and coordinates the equipment in the local control layer, to realize control over the grid-connected operation and the grid-disconnected operation of the microgrid; the local control layer consists of distributed resources, energy storages, load controller and intelligent terminal, and other devices, which can realize data collection, local protection control, distributed generation regulation, energy storage charging/discharging control, and load control. For a small-sized

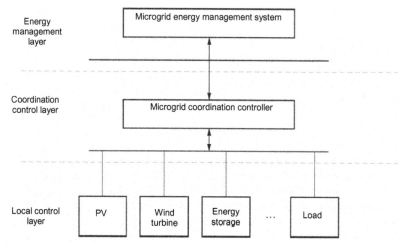

FIG. 3.46   Structure microgrid hierarchical control.

simple microgrid, the structure can be simplified, and a two-layer control structure by combining the energy management layer with the coordinate control layer can be adopted.

3.7.1.1   Grid-connected operation

In grid-connected operation of the microgrid, the microgrid is connected to the grid via PCC. MGCC is responsible for the management over the main energy storage battery, to control the SOC at the upper limit to store as much energy as possible, so that the microgrid can operate in the grid-disconnected mode as long as possible.

3.7.1.2   Grid connection to disconnection

In case of external power source loss, the microgrid needs to shift from the grid-connected mode to the grid disconnected mode (unplanned islanding). In case of planned scheduling, the microgrid needs to shift to the grid disconnected mode (planned islanding).

(1) PCC switching power regulation for planned islanding

In case of planned islanding, MGCC will, regulate the output of the energy storage according to the scheduling instruction, to control the switching power at PCC to be 0, and maintain the power balance among the energy storage, DG output, and load. Then MGCC will send PCC trip instruction, energy storage will shift from the P/Q mode to the V/f mode, and the microgrid shift to the grid-disconnected mode. The purpose of MGCC regulating the output of the energy storage and controlling the switching power at PCC to be zero is to avoid over-large conversion power. If the switching happens when the microgrid generates a large amount of power or when the microgrid delivers large active power to the grid, it may result in transient overvoltage and trigger overvoltage protection of the main energy storage during the microgrid is disconnected from the grid.

(2) Overvoltage during unplanned islanding

During unplanned islanding, if power amount generated by the microgrid is large or the active power delivered to the grid is large, the sudden energy unbalance on the microgrid will result in microgrid switching failure and generates islanding protection overvoltage, which will cause main energy storage shutdown and further result in grid disconnection failure.

(3) "Gap" during unplanned islanding

During unplanned islanding, if the switching power at PPC is not large enough, the islanding overvoltage will not result in main energy storage shutdown. After islanding is identified, the PPC switch will be disconnected and shift to the grid disconnected operation mode. During this process, firstly, power loss will occur on the microgrid, which will result in voltage and frequency fluctuation on the microgrid. When the system is disconnected from the grid, the main energy storage will shift from the P/Q control mode to the V/f control mode, and recover the voltage and frequency on the microgrid to the normal state. The period from sudden power loss to normal state recovery is called "gap" for unplanned islanding. Although the quick electronic switch can narrow the "gap," it is hard to fully eliminate. This "gap" problem remains to be solved in the microgrid sector.

3.7.1.3 Grid-disconnected operation

(1) Battery charging/discharging management

During grid-disconnected operation, MGCC will perform active management over the battery of the main energy store, so as to maximize the utilization of distributed generation, and ensure the microgrid can operate in the grid disconnected mode as long as possible. In case of small load, MGCC will be responsible for energy storage management, to store surplus energy from DG. If the power content of the main energy storage reaches up to the upper limit, MGCC will restrain the charging of the main energy storage. Otherwise, overcharge protection will be triggered and result in shutdown.

(2) Distributed generation control

During grid-disconnected operation of the microgrid, MGCC will predict the distributed generation and the loads condition, and control the output of each DG and energy storage according to the current, voltage, power, and switching value collected at each node, so as to realize off-grid energy balance of the microgrid. In case of MGCC communication failure, the control over the distributed generation cannot be realized. In this case, the microgrid cannot operate normally.

3.7.1.4 Grid disconnection to connection

During the grid-disconnected operation of the microgrid, if the power source of the grid recovers to the normal state or the microgrid resumes from the unplanned islanding, the microgrid will be reconnected to the grid for operation. Due to the angle and frequency differences existing between the voltage of the microgrid under grid-disconnected mode and the voltage of the distribution grid, a synchronization check will be performed before grid connection. The microgrid will be reconnected to the grid after the angle difference falls within the limit, so as to reduce reconnecting surge generated at the moment of reconnection. If the connecting surge is large, the surge current will trigger the overcurrent protection of the main energy storage and cause the shutdown of the energy storage, which will result in the power outage of the whole microgrid. Therefore, the connecting surge will be reduced as much as possible when reconnecting the microgrid to the grid, so as to realize smooth shifting.

According to the previous analysis, MGCC is an essential part of grid-connected operation, connection-to-disconnection, grid-disconnected operation, and disconnection-to-connection, and MGCC highly relies on communication. Besides, it cannot solve problems such as overvoltage, switching "gap," and connecting surge.

3.7.2 Frequency-shift control technology

Frequency-Shift-Keying technology refers to a technology of modulating carrier frequency with digital signals, which is a mature technology applied in communication field for power system protection, such as when the FSK transceiver is used for high voltage line protection, with a rated frequency range of 50–400 kHz and rated bandwidth of 4 kHz. Under normal operation, it will send frequency monitoring signals with a frequency of f_G, which is for channel monitoring. In case of failure, it will send command signals with a frequency of f_T, which is for sending operating instructions. The carrier terminal for high-voltage line protection adopts FSK. Five frequencies are provided and can be shifted in one channel. In normal operation condition, it will send frequency monitoring signal f_G. In case of failure, it will send frequency hopping f_A, f_B, f_C, and f_3. Frequency hopping f_A, f_B, and f_C correspond to Phase A, Phase B, and Phase C, respectively, and f_3 is the frequency hopping for three-phase tripping, as shown in Fig. 3.47.

The microgrid can borrow the idea of FSK technology. During grid-disconnected operation of the microgrid, with power frequency signal of the voltage source, by adopting the Frequency-Shift-Keying technology and by taking the frequency signals as the means of communication, control over the communication line–free interconnected microgrid can be realized. In this method, MGCC is not required, which is a "plug-and-play" microgrid control way with the simplest physical structure.

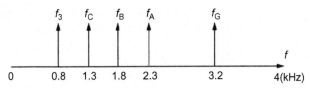

FIG. 3.47 Frequency-shift-keying modulation.

3.7.2.1 Frequency partition

As shown in Fig. 3.48, referring to the requirements for frequency deviation fault ride-through of the 0.5–100 mW generator set, for the range of 47.5–51 Hz, the range of 49 Hz to 50.3 Hz is the normal operating frequency range of the generator set; 50.3–51 Hz is the frequency range falling beyond the upper limit. Under this condition, the allowable frequency deviation fault ride-through time is 30 min, and 47.5–49 Hz is the frequency range falling beyond the lower limit; under this condition, the allowable frequency deviation fault ride-through time is 30 min. During grid-disconnected operation of the microgrid, the main energy storage adopts the virtual synchronous generator technology, which possesses the external characteristics of the voltage source. With the power frequency of the voltage source, adjust the lower frequency limit from 47.5 Hz to 47.7H. During charging/discharging of the main energy storage under the maximum power, as shown in Fig. 3.49, the frequency fall within the range of 47.7–51 Hz. No matter how severe the power fluctuation is, so long as it not exceeding the maximum charging and discharging power, the system

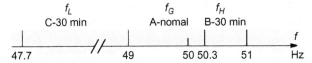

FIG. 3.48 Frequency partition.

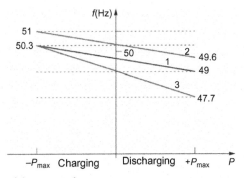

FIG. 3.49 SOC-based droop control.

frequency will not exceed the allowable range, which can control the voltage and frequency stability simultaneously, and has strong robustness. In detail, divide the frequency range into three sections according to SOC status, 50.3–51 Hz is the high SOC and charging droop section and also the high frequency fault ride-through section; 47.7–49 Hz is the low SOC discharging droop section and also the low frequency fault ride-through section; 49–50.3 Hz is the normal SOC droop section and the high SOC discharging/low SOC charging droop area. SOC sends monitoring frequency f_G, high SOC charging sends high control frequency f_H (control distributed generation), and low SOC discharging sends low control frequency f_L (under frequency load shedding and load cutting).

The SOC-based droop control method is shown in Fig. 3.49. In the grid-connected operation mode, Line 1 in Fig. 3.49 is the main energy storage operating droop line. The normal operating frequency range f_G is 49–50.3 Hz. The main energy storage will maintain the energy storage battery according to the condition of SOC, so that SOC can fall within the normal range.

During grid-disconnected operation, in case of normal SOC, the operating droop line of the main energy storage is the Line 1 in Fig. 3.49, and the droop operating frequency range f_G is between 49 and 50.3 Hz. The main energy storage maintains the energy storage battery according to the SOC status, so as to ensure SOC falls within the normal range. In case of high SOC, the main energy storage droop line is the Line 2 in Fig. 3.49, and the droop operating frequency is between 49.6 and 51 Hz, in which f_H falls within 50.3–51 Hz. During the charging process, the main energy storage droop operates in this section, sending charging prohibited signals, and the power generation of the generating unit of DG is regulated according to such frequency signal. In addition, the power amount generated by DG is subject to the regulation by the high frequency, and the main energy storage will be shifted out from the charging stage. During the discharging process, the main energy storage droop operating range falls within the 50.3 Hz to 49.6 Hz range, regulation on the power generation of the DG generating unit will be stopped, and along with battery discharging, SOC will resume to the normal state. In case of low SOC, the main energy storage droop line is the Line 3 in Fig. 3.49, the droop operating frequency falls within the 47.7–50.3 Hz range, in which f_L is between 47.7 and 49 Hz. During the discharging process, the main energy storage droop operates in this section. Due to low SOC, discharging is not allowed. The signal at this frequency will be used as the underfrequency load shedding signal. Through underfrequency load shedding, ensure the main energy storage works at a frequency higher than 49 Hz; realize battery charging. Send signals at different frequencies within the allowable frequency range, and regulate the power generation of the distributed resources and loads according to such signals, so as to realize control over the communication line-free interconnected microgrid.

3.7.2.2   Main energy storage droop control line

According to the previously mentioned main energy storage droop line, the Droop characteristics of the output active power and frequency are as follows:

$$f = f_n + k_i P + C_i \tag{3.40}$$

where f stands for the frequency of the output power source of the main energy storage; f_n stands for the rated frequency of the power source; k_i stands for the droop coefficient; P stands for the output active power; P_{max} stands for the maximum output active power of the main energy storage; the maximum absorbed active power $P_{min} = -P_{max}$; and C_i is a constant.

According to the SOC status of the main energy storage, different droop lines will be used to indicate normal SOC, high SOC, and low SOC.

(1) Normal SOC

Consider line 1 in Fig. 3.49. When the upper limit of the frequency $f_{max} = 50.3\,\text{Hz}$, $P = -P_{max}$. When the lower limit of the frequency $f_{min} = 49\,\text{Hz}$, $P = P_{max}$.
We can obtain.

$$k_i = -(f_{max} - f_{min})/(P_{max} - P_{min}) = -0.65/P_{max}$$

Take $C_i = -0.35$.
Substitute k_i, C_i into Formula (3.40):

$$f = f_n - 0.65P/P_{max} - 0.35 \tag{3.41}$$

The physical meaning of Formula (3.41) is that, in case of normal SOC, the droop line f can fall within the allowable frequency range from 49 to 50.3 Hz, which is the f_G range. During the discharging process, when outputting the maximum active power P_{max}, the frequency will reach the allowable lower limit, 49 Hz; during the charging process, when absorbing the maximum active power P_{max}, the frequency will reach to the allowable upper limit 50.3 Hz. During the noncharging and nondischarging period when $P = 0$, the frequency falls to the intermediate value $f_0 = 49.65\,\text{Hz}$, as shown in Line 1 in Fig. 3.49, which corresponds to f_G in Fig. 3.48.

(2) High SOC

Consider the line 2 in Fig. 3.49. When the upper limit of the frequency $f_{max} = 51$ HZ, $P = -P_{max}$. During the period without charging and discharging $P = 0$, then $f_0 = 50.3\,\text{Hz}$; we can obtain

$$k_i = -(f_{max} - f_0)/P_{max} = -0.7/P_{max}$$

Take $C_i = 0.3$; substitute into the formula to obtain

$$f = f_n - 0.7P/P_{max} + 0.3 \tag{3.42}$$

The physical meaning of Formula (3.42) is as follows: in case of high SOC, during the charging process, the droop line f falls within in the range of 50.3–51 Hz, which belongs to f_H section. During the charging process $P < 0$, entering the f_H section, DG generation control signals are sent out. During the charging process, when absorbing the maximum active power P_{max}, the upper limit of f_H (i.e., 51 Hz) is reached, and thee frequency signals are used for regulating power generation of generating unit of DG, as shown in Line 2 in Fig. 3.49, which corresponds to f_H in Fig. 3.48. During the discharging process, $P > 0$, f falls within 49.6–50.3 Hz, which belongs to the f_G section.

(3) Low SOC

Consider line 3 in Fig. 3.49. When the lower limit of the frequency $f_{min} = 47.7$ Hz, $P = P_{max}$, and when the upper limit of the frequency $f_{max} = 50.3$ Hz, $P = -P_{max}$, we can obtain:

$$k_i = -(f_{max} - f_{min})/(P_{max} - P_{min}) = -1.3/P_{max}$$

Take $C_i = -1$ and substitute into the formula to obtain:

$$f = f_n - 1.3P/P_{max} - 1 \tag{3.43}$$

Physical meaning of Formula (3.43) is as follows: in case of low SOC, during the discharging process, the droop line f falls within the range of 47.7–49 Hz, which belongs to the f_L section. During the discharging process, $P > 0$, entering the f_L section, the underfrequency load shedding and load cutting instructions are sent out. During the discharging process, when outputting the maximum active power P_{max}, the lower limit of f_L (i.e., 47.7 Hz) is reached, and this frequency signal serves as the underfrequency load shedding, as shown in Line 3 in Fig. 3.49, which corresponds to f_L in Fig. 3.48. During the charging process, $P < 0$, f falls within 49–50.3 Hz, which belongs to the f_G section.

3.7.2.3 Auto overfrequency/power control (f/P) of DG

During the microgrid is in grid-disconnected operation mode, the off-grid energy balance on the microgrid will be maintained in a real-time manner. MPPT technology is adopted, so as to improve the availability of DG as much as possible. In case of high battery SOC, the excess energy can neither be stored nor consumed by loads, and overlarge amount of energy will lead to overvoltage, which can result in off-grid energy unbalance on the microgrid and cause microgrid failure. In this case, DG output will be controlled and regulated so as to ensure off-grid energy balance.

During grid-disconnected operation, in case of high SOC, the main energy storage works as per Line 2. The droop line undergoes the charging process. The charging process falls within the f_H section. In this case, power amount generated by DG will be regulated with frequency (see Section 3.3 for details).

3.7.2.4 Main energy storage V/f control

During the grid-disconnected operation, the problem of off-grid energy balance exists. During the grid-disconnected operation, the main energy storage adopts the Frequency-Shift-Keying technology, sending different droop frequencies according to the battery SOC conditions, and DG will then regulate its output according to such frequencies, and ensure off-grid energy balance. During the grid-connected operation, the energy storage converter works in the P/Q mode, and the grid frequency is determined by the microgrid. The energy storage works in the P/Q mode during the grid-connected operation, and in the V/f mode during the grid-disconnected operation. Therefore mode shifting is necessary for connection-to-disconnection switching, and such shifting will bring about the problem of switching "gap."

The virtual synchronous generator possesses the external characteristics of a voltage source (see Section 3.5 for detail), which has two operating modes: the grid-connected mode and the grid-disconnected mode. Therefore, during planned islanding or unplanned islanding, it maintains in the initial state of grid-connected operation, so as to realize seamless shifting from the grid-connected mode to the grid disconnected mode. The presynchronous grid-connected technology can be adopted to solve the closing surge problem during the connection-to-disconnection switching process. Refer to Section 3.4 for details. When the connection-to-disconnection instruction is received, monitor the voltage phase information at PCC at grid side in a real-time manner, and regulate the voltage phase of the energy storage converter according to the phase regulating step setting until the voltage phases at the two sides are synchronous. Then send PCC closing instruction for grid connection, so as to ensure surge free closing during the disconnection-to-connection switching process.

3.7.2.5 Battery charging/discharging management

During the grid-connected operation, the scheduling end of the distribution grid will regulate the power, so as to control the charging and discharging state of the energy storage, and maintain the same in the rational SOC state. During the grid-disconnected operation, the power to loads should be supplied by DG, rather than the energy storage as far as possible. In addition, surplus energy will be stored in the battery, so as to prolong the grid-disconnected operation. The upper limit of energy storage SOC will be set to 80%–90%, and the lower limit to 20%–30%. During the grid-connected operation, the energy storage SOC will fall within the upper limit range at about 80–90%, so as to ensure long-term power supply by the energy storage during grid-disconnected operation.

During the grid-disconnected operation, when SOC fall beyond the upper limit, the energy storage works as per Line 2 in Fig. 3.49, and the charging process falls within the f_H section, sending the charging prohibited signal. In case of normal SOC, the energy storage works along the Line 1 in Fig. 3.49, sending the charging/discharging allowed signal.

FIG. 3.50 Relations between SOC and battery charging and discharging.

The energy storage SOC may decrease along with load consumption. When SOC falls beyond the lower limit, the energy storage works along the Line 3 in Fig. 3.49, and the charging process falls within the f_L section, sending the discharging prohibited signal. Relations between SOC and battery charging/discharging status are as shown in Fig. 3.50.

3.7.3 Experimental verification

Fig. 3.51 shows the diagram of communication line-free microgrid experiment system. The experiment system consists of a 20 kW energy storage, a 20 kW load, and a 20 kW PV generator. The energy storage battery is of lithium-ion battery. The energy storage converter adopts the virtual synchronous generator technology, which possesses the external characteristics of the

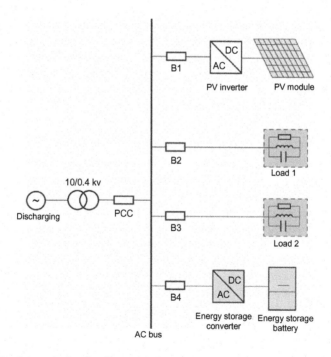

FIG. 3.51 Communication line-free microgrid experiment system.

voltage source. The droop line adopts the frequency shift keying technology proposed in this book.

Presynchronous grid connection technology is adopted. The PV inverter adopts f/P line control technology. The power of Load 1 is 12 kW, the power of Load 2 is 8 kW, the AC bus voltage supply to the experiment system is 400 V, and MGCC is not provided in this system. The experiment items are as follows: the connection-to-disconnection switching experiment, the experiment of load sudden change during grid-disconnected operation, the disconnection-to-connection switching experiment, and so forth, so as to verify the control effect of the communication line-free interconnected microgrid under different working conditions.

3.7.3.1 Connection-to-disconnection switching experiment

The objective of this experiment is to verify that the control system is able to solve the overvoltage on the microgrid at the disconnecting moment and the "gap" switching problem during the connection-to disconnection switching process. Fig. 3.52 shows the connection-to-disconnection switching experiment. In the experiment, power is supplied under the maximum conversion power from the microgrid to the distribution grid. Power generated by PV generation is 20 kW; and power consumption by loads is 0 kW. During grid-connected operation, the 20 kW power generated by PV generation is delivered to the distribution grid via PCC, and no power is output from the energy storage. In this case, unplanned islanding occurs, and 20 kW power generated by PV generation is fully absorbed by the energy storage. In the figure, Line 2 indicates the bus voltage (390 V) of the microgrid. Seamless connection-to-disconnection switching is realized, and no overvoltage occurs, which realize seamless switching during unplanned islanding. Line 3 indicates the current of the energy storage converter, the output of the energy storage changes from 0 to the charging current 29.8A when islanding occurs.

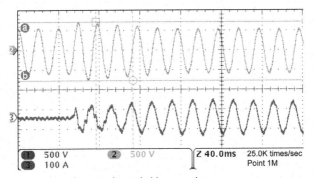

FIG. 3.52 Connection-to-disconnection switching experiment.

3.7.3.2 Experiment of sudden load change in the grid-disconnected mode

The objective of this experiment is to verify that during the grid-disconnected operation, the control regulation and energy balance can be realized merely by the coordination between the main energy storage and the distributed resources without relying on communication. Fig. 3.53 shows the experiment of load sudden change during the grid-disconnected operation. Under normal operation, the loads are 20 kW, while the output of PV generation is 10 kW, and output of energy storage is 10 kW. According to the figure, the load suddenly reduced by 12 kW at 90 ms, and is suddenly reduced by 8 kW at 270 ms. After first load reduction (reduced by 12 kW, 8 KW remains), the power generated by PV generation is 10 kW, which exceeds the load demands. The redundant 2 kW is absorbed by the energy storage, and the energy storage shifts from the discharging state to the charging state. After the second load reduction, the load is reduced from 8 kW to 0. The power generated by PV generation is 10 kW and the load consumption is 0. Therefore, the 10 kW power generated by PV generation is fully absorbed by the energy storage. In the figure, Line 1 indicates the voltage of microgrid bus is 390 V, which remains unchanged; Line 3 indicates the current at the AC side of the energy storage, the current phase difference before and after load reduction is 180 degrees, and current shifts from discharging to charging (0–90 ms: 14.9 A, 90–270 ms: 3.1 A, after 270 ms: −15.2 A); Line 2 indicates the voltage (700 V) at the DC side of the energy storage converter; and Line 4 indicates the current at the DC side of the storage energy converter (0–90 ms: 14.1 A, 90–270 ms: 2.9 A, after 270 ms: 14.5 A).

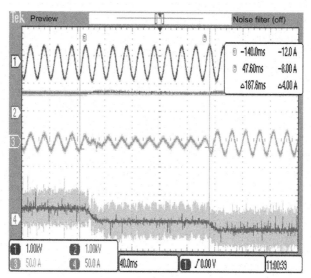

FIG. 3.53 Experiment of sudden load change in the grid-disconnected mode.

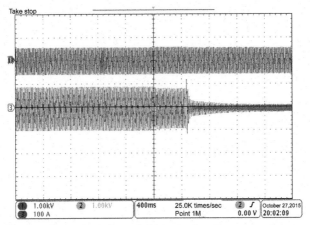

Take stop

| 1.00kV | 2 1.00kV | 400ms | 25.0K times/sec | 2 ∫ | October 27,2015 |
| 100 A | | | Point 1M | 0.00 V | 20:02:09 |

FIG. 3.54 Disconnection-to-connection switching experiment.

3.7.3.3 Disconnection-to-connection switching experiment

Fig. 3.54 shows the experiment of disconnection-to-connection switching after unplanned islanding. The microgrid operates in the grid-disconnected mode with 20 kW loads connected. Power to the loads is fully supplied by the energy storage. When the voltage is detected at distribution grid side, PCC needs to be closed, then the output of the energy storage decreases to 0, and power to loads are supplied by the distributed grid. No surge occurs to the energy storage during the disconnection-to-connection switching to realize smooth switching. In the figure, Line 1 indicates the voltage of microgrid bus, which remains unchanged at 390 V; Line 3 indicates the current at AC side of the energy storage, and current change from 29.6 A to 0.

Chapter 4

Protection and control technologies of connecting to the grid for distributed power resources

The traditional power distribution network receives the electric energy from the power transmission network and distributes it to the consumers through the power distribution facilities. The current flows unidirectionally without active control. Hence, it is called the *passive distribution network (PDN)*. The current flows bidirectionally when the distributed resources are connected into the traditional power distribution network, influencing the traditional power distribution network in terms of the short circuit level, equipment selection, reactive power, voltage distribution, distribution network protection, distribution automation, fault handling, and island operation under special conditions. The traditional power distribution network is unsuitable for connection of various distributed resources because it is not specially designed for the distributed resources. The active distribution network (ADN) technology was put forward by the Power Distribution and Distributed Generation Committee of the International Council on Large Electric Systems (CIGRE) in 2008. ADN is a distribution network, which manages the current flow with a flexible topology structure, in order to facilitate active control and active management of the local distributed resources. This chapter introduces the technologies surrounding grid connection of the distributed resources, protection, and control of the distribution network.

4.1 Pilot protection for overhead active distribution network

The feeder protection of the one-way radial nonautomatic power distribution network adopts the three-section current protection. The influences on protection of the distribution network caused by connection of the distributed resources mainly include decrease of the protection sensitivity due to the fault

Distributed Power Resources. https://doi.org/10.1016/B978-0-12-817447-0.00004-3

current infeed at the end, false action of the protection due to the fault adjacent lines, and failure of reclosing. The distribution network feeder protection can be modified with the directional element added, in order to deal with the influences caused by connection of the distributed resources by virtue of the directional current-voltage protection.

In the automatic distribution network, different feeder automation (FA) protection programs are adopted based on the connection modes of the distribution network for distribution automation (DA). There are two connection modes in the distribution network: the overhead network connection and the cable network connection. This section introduces the protection program of the overhead active distribution network.

4.1.1 Influences on overhead distribution network caused by the connection of distributed resources

Fig. 4.1 shows the four network structures for connection of the overhead distribution network: the single power supply radial network, the dual power supply hand-in-hand ring network, the segmented two-link network, and the segmented three-link network. The switches include the recloser, the sectionalizer, and the coupling breaker. FA programs of the overhead distribution network include two types: the distributed FA and the centralized FA.

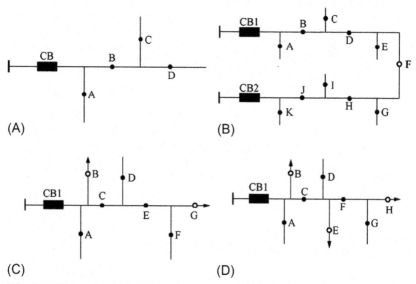

FIG. 4.1 Wiring of overhead distribution network. (A) Single power supply radial network; (B) dual power supply hand-in-hand ring network; (C) segmented two-link network; and (D) segmented three-link network. ■, Recloser; ●, Sectionalizer; ○, Coupling breaker.

4.1.1.1 Distributed FA of overhead network

The distribution protection of distributed FA is realized in the in-place manner, with the current protection, the recloser, and the sectionalizer equipped for fault isolation. The protection is classified into the voltage-time type, the current counting type, and the current-voltage hybrid type based on the model of sectionalizer selected. Fig. 4.2 shows the typical ring network voltage-time type fault handling process. The process of the current counting type and the current-voltage hybrid type is similar to that of the voltage-time type shown in this diagram.

In Fig. 4.2, sectionalization and coupling of the overhead line is realized by the circuit breaker and the pole-mounted load switch. The recloser realizes current protection, circuit breaking, and closing via the circuit breaker. The sectionalizer realizes automatic breaking without voltage via the pole-mounted load switch, with the wires connected in form of the typical hand-in-hand ring network. The fault handling process is as follows: CB1 trips when a fault occurs. The sectionalizers B and C break automatically without voltage after

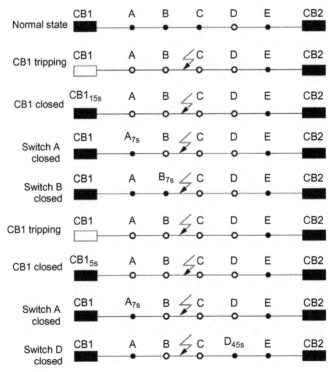

FIG. 4.2 Fault handling of typical voltage-time type ring network. ■, Recloser; ●, Sectionalizer.

CB1 is reclosed. The sectionalizers B and C close automatically with voltage supplied. The fault section is identified. Closing is performed several times for fault isolation.

The fault isolation time will be 1–2 min generally.

The FA protection program allows for ideal fault isolation without the DG connected. In case of DG being connected to B or C, when the power supply outside CB1 section fails, DG will supply fault current, which will pass through CB1 and cause incorrect action of CB1 of multiple reclosing; for DG with the function of island detection, in case of a fault inside the zone, after CB1 trips, the automatic breaking without voltage of the sectionalizers B and C can be realized only after the island protection of DG acts (the longest detection time 2 s) due to the existence of DG and voltage in the sectionalizers B and C. Automatic breaking is impossible if the island protection of DG fails to act. For DG with low voltage ride through capability, the action time is set as 3 s when the fault voltage is larger than $0.9U_\mathrm{N}$, and the sectionalizers B and C break more slowly. If the fault voltage is not below $0.9U_\mathrm{N}$, the sectionalizers B and C fail to realize automatic breaking without voltage.

4.1.1.2 Centralized FA of overhead network

In the centralized FA, the distribution protection is realized by the main station (substation) and the feeder terminal unit (FTU), with the fault section identified. Take Fig. 4.2 as an example. Both the recloser and sectionalizer are installed with FTU, which is connected to the main station (substation) via the communication devices.

Without DG connected, in case of a fault in the section, CB1 trips due to the overcurrent and successfully completes instantaneous fault reclosing. When the permanent fault reclosing fails, the main station (substation) fault handling is started; FTU sends the fault information to the main station (substation). The main station (substation) identifies the inconsistency of the overcurrent state of the neighboring FTUs, thus judging that the fault occurs in different FTUs. The main station (substation) remotely controls the corresponding FTU to break for isolating the fault section.

With DG connected, the fault can also occur at the side of power supply outside the CB1 section, which can result in fault current supplied from DG. CB1 will act incorrectly when the fault current flows through CB1. In case of an instantaneous fault inside the section, after CB1 trips, reclosing can fail due to the failure of arc extinguishing because of DG. In case of the permanent reclosing failure, the main station (substation) will be started for fault handling. FTU will send the fault information to the main station (substation). The main station (substation) judges that the overcurrent states of the neighboring FTUs are consistent due to DG, which can result in the failure of fault section recognition and fault isolation.

4.1.2   Pilot protection of multibranch line

4.1.2.1   Pilot protection of three-branch line

For the traditional high voltage line, the blocking mode pilot protection and the transceiver are used for protection of the high voltage branch line. This is a mature engineering application program for the high voltage branch line relay protection before the optical pilot three-terminal current differential protection occurs. Fig. 4.3 shows the pilot relay protection of the branch line with power supply at all three sides.

When $F1$ inside the section fails, I_{S_1}, I_{S_2}, and I_{S_3} of the starting elements S_1, S_2, and S_3 are larger than the fixed value of starting I_{QD}; the positive direction elements of S_1, S_2, and S_3 will act ($F_{S_1}{}^+ = 1, F_{S_2}{}^+ = 1, F_{S_3}{}^+ = 1$). Hence, it can be identified as a fault inside the section. Therefore, the start criterion is

$$I_{S_1} > I_{QD}, I_{S_2} > I_{QD}, I_{S_3} > I_{QD} \tag{4.1}$$

The action criterion is

$$F_{S_1}^+ = 1, F_{S_2}^+ = 1, F_{S_3}^+ = 1 \tag{4.2}$$

When $F2$ outside the section fails, the starting elements S_1, S_2, and S_3 (I_{S_1}, I_{S_2}, and I_{S_3}) will act ($F_{S_1}^+ = 1, F_{S_2}^+ = 1$); the positive direction elements of S_1 and S_2 will act ($F_{S_1}^+ = 1, F_{S_2}^+ = 1,$), and the positive direction element of S_3 does not act ($F_{S_3}^+ \neq 1$). This is inconsistent with the expression (4.2). Hence, it can be identified as a fault outside the section.

4.1.2.2   Pilot protection of multibranch line

The pilot protection principle applies when the pilot protection of multibranch line expands from the three branch structure to the multi branch structure. Fig. 4.4 shows the pilot protection of multibranch lines with power supply.

When $F1$ inside the section fails, I_{S_1}, I_{S_2},... I_{S_n} of the starting elements S_1, S_2,... S_n are larger than the fixed value of starting I_{QD}, the positive direction elements of S_1, S_2,... S_n will act ($F_{S_1}^+ = 1, F_{S_2}^+ = 1, F_{S_3}^+ = 1,... F_{S_n}^+ = 1$). Hence, it can be identified as a fault inside the section. Therefore, the start criterion is

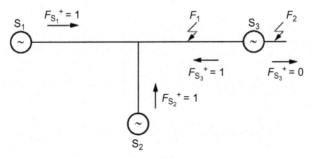

FIG. 4.3   Pilot protection of active three-branch line.

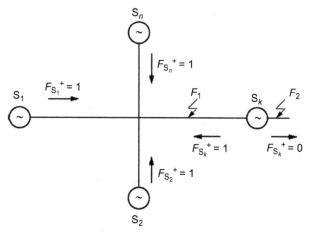

FIG. 4.4 Pilot protection of multibranch line with power supply.

$$I_{S_1} > I_{QD}, I_{S_2} > I_{QD}, \dots I_{S_n} > I_{QD} \tag{4.3}$$

The action criterion is

$$F_{S_1^+} = 1, F_{S_2^+} = 1, \dots F_{S_n^+} = 1 \tag{4.4}$$

When $F2$ outside the section of K branch fails, the starting elements of S_1, $S_2, \dots S_n$ will act ($I_{S_1}, I_{S_2}, \dots I_{S_n}$).

The positive direction element of S_k does not act ($F_{S_k^+} \neq 1$), and those of all the other branches act. This is inconsistent with the expression (4.4). Hence, it can be identified as a fault outside the section.

4.1.2.3 Pilot protection without power supply of multibrance line

The expressions (4.3) and (4.4) apply to the multibranch lines with power supply. If a line has no power supply, the line fails, the branch will not supply the short circuit current, and the device will not start. Hence, the directional element will not be used for comparison. The expressions (4.3) and (4.4) need to be modified for the multibranches without power supply, as shown in Fig. 4.5. The expression is

The start criterion is

$$I_{S_i} > I_{QD}, I_{S_k} < I_{QD} \quad i \geq 1, k \leq n \tag{4.5}$$

The action criterion is

$$\Pi F_{S_i^+} = 1 \quad i \neq k \tag{4.6}$$

wherein I_{S_i} is the staring element of the branch i, and $F_{S_i^+}$ is the directional element of the branch i.

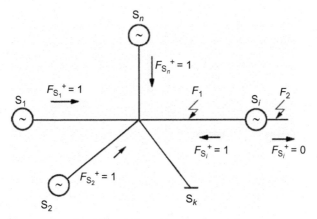

FIG. 4.5 Pilot protection of multibranch line without power supply.

The physical meaning of the expression: In case of a fault inside the section, the starting element of the branch with power supply will act, the branch without power supply will not act, and all the positive direction elements in the branches with power supply, which are excluded for action judgment, and will act ($F_{S_i^+} = 1$). Hence, it can be identified as a fault inside the section. In case of a fault outside the section, the starting element of the branch with power supply I_{S_i} will act; I_{S_k} of the branch without power supply, which is excluded for action judgment, and will not act. The positive direction element of the branch with power supply S_i (back fault) will not act ($F_{S_i^+} \neq 1$), and all the positive direction elements of the other branches with power supply will act. This is inconsistent with the formula (4.6). It can be identified as a fault outside the section.

4.1.3 Pilot protection for overhead active distribution network based on virtual node network topology

4.1.3.1 Classification of virtual nodes for network topology of overhead distribution network

To classify the topology structure of the distribution network in Fig. 4.1, the recloser, the sectionalizer, and the coupling breaker are classified as the modular elements. With the line simulated as a node, classify the unit into different branch sections by the node, and use the pilot protection of multibranch lines without power supply as the criterion of judgment (Formula (4.5) and Formula (4.6)), in order to realize the pilot protection of the distribution network. Take the radial network wiring with single power supply and the segmented three-link wiring structures in Fig. 4.6 as examples for sectioning of the different node unit protection sections. As shown in the diagram, the single power supply radial network wire can be divided into two nodes. Each node has three branches. The connection wire of Section 3 can be divided into three nodes.

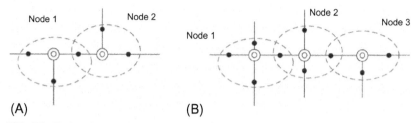

FIG. 4.6 Node unit protection sections. (A) Wiring of single power supply radial network; (B) wiring of segmented three link network. ● Unit element; ◎ Virtual node.

Node 1 and Node 2 have four branches, and Node 3 has three branches. This method applies to all the distribution networks. Different nodes can be applied for different distribution network structures. Each node realizes the multi-branch pilot protection for the active distribution network.

4.1.3.2 Pilot protection of overhead distribution network with EPON communication

The Ethernet passive optical network (EPON) features the single point to multiple point structure and applies to the automatic communication program for overhead distribution. EPON does not support the communication between the optical network units (ONU). The single fiber wave length time sharing multiplexing technology, with downlink wavelength of 1490 nm and uplink wavelength of 1310 nm, is adopted for communication between the optical line terminal (OLT) and multiple ONUs.

The upward and downward signals of different wavelengths are used in one optical fiber. It is impossible to realize synchronized sampling based on fiber channel due to the inconsistent upward and downward transmission delay. Hence, it is improper to adopt the current differential protection program based on synchronization for distribution automation with EPON communication. However, the branch pilot protection program has no requirement for synchronization, and can use the generic object oriented substation event (GOOSE) mechanism based on EPON. FTU transmits the starting and direction information to the centralized pilot protection via GOOSE. The centralized pilot protection adopts the pilot distribution network protection based on the virtual node network topology structure. Hence, it is ideal for protection of DG connected into ADN.

The distributed resources feature intermittency (e.g., the PV system generates electricity at daytime and does not work at nighttime), and it generates electricity on the sunny days, and does not work on the rainy days; the wind power system generates electricity on the windy days and does not work on the windless days. Hence, the power supply can be available or unavailable when the branch of distributed resources is connected. The distribution network pilot protection program based on the virtual node network topology structure is well adapted to the distributed resources. When no electricity is generated in the branch of distributed resources system, this branch will be excluded from judgment automatically and works as the normal load branch.

The unit elements (recloser, sectionalizer, and coupling breaker) in the distribution network are equipped with corresponding FTUs (built-in ONU). FTU collects the current and voltage information at site, the starting and direction elements act based on Formula (4.5) and Formula (4.6). ONU transmits the information collected to OLT via the GOOSE transmission mechanism of EPON communication. OLT is connected with the centralized pilot protection, which is arranged with multiple node protection sections based on the distribution network topology structure, to realize the protection of the active distribution network. Fig. 4.7 shows the centralized pilot protection based on EPON communication.

4.1.4 Test verification

Fig. 4.8 shows the wiring diagram for system test. The testing devices include the IDP831 distribution network protection feeder terminal unit (FTU), the built-in ONU, the IDP801 centralized pilot protection device, the network switch, ONU, the optical splitter, OLT, and the network message analyzer.

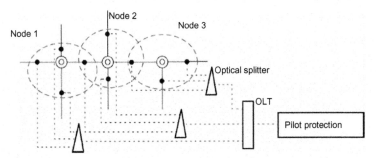

FIG. 4.7 Centralized pilot protection based on EPON communication.

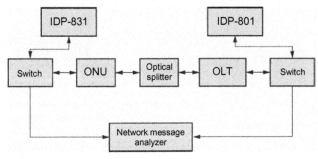

FIG. 4.8 Wiring diagram for system test.

The IDP831 distribution network protection feeder terminal unit uploads the information of starting and direction elements, and transmits it to the IDP801 centralized pilot protection device via the network consisting of the network switch, ONU, the optical splitter, and OLT. The centralized pilot protection adopts the pilot distribution network protection based on the virtual node network topology structure.

GOOSE mechanism of EPON transmission adopts the time sharing multiplexing mode for the uplink signals; the time is long and adopts the broadcast mode for the downlink signals, with a short time delay. The uplink time delay of GOOSE is 1.34 ms, and the downlink time delay of GOOSE is 0.04 ms in the actual test, which meet the requirements of the channel for distribution network protection.

Fig. 4.9 shows the action waveform actually tested. In the test, *V/V* wiring (Phase B earthed, the voltage is zero) is adopted for voltage testing, and Phase A and Phase C connection are adopted for current testing. In view that the requirement for the action time of distribution network protection is not so high as that of the high voltage line protection, the 45 ms power backward direction delay and GOOSE shake elimination delay is added for protection judgment of the centralized protection device, respectively. In the actual test, the action time of the overcurrent starting element and the direction element is 23 ms, and that of the whole unit protection is 70 ms (receiving GOOSE tripping time).

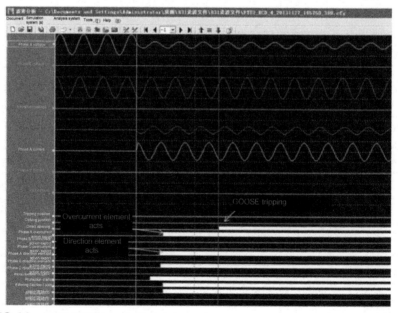

FIG. 4.9 Action waveform actually tested.

4.1.5   Application case

Fig. 4.10 is the wiring of the distribution network in Nanji Island microgrid demonstration project of the State 863 Program, which is dominated by the overhead lines and adopts the single power supply radial network. The outgoing line of power supply is connected to the 1 MW wind power generation system, the 435 kW PV power generation system, and the radial branch of the distribution network is connected to the 110 kW PV power generation system. The power of the main power supply for the diesel engine of the microgrid is 1.6 MW. The distribution network structure includes two single power supply radial parts (in the blocks in the diagram) and adopts EPON communication. FTU and a series of centralized pilot protection devices realize protection of the active distribution network via EPON communication.

This project is a 10 kV distribution network. The original structure is of the overhead single radial connection type. The original distribution network protection program adopts the current protection, the recloser, and the sectionalizer to realize fault isolation. The wiring diagram is as shown in Fig. 4.11.

In case of interphase short circuit in Section D of trunk main, the handling process is as shown in Fig. 4.12.

In this program, reclosing will be conducted repeatedly for fault isolation which takes 25 s. In case of the distributed resources connected in Section C, after the sectional recloser FB trips, there is still voltage existing in the voltage type sectional load switches FI1 and FI2, which results in the failure of normal breaking and influences fault positioning and isolation. Hence, this program fails to meet the demands of the distribution network in terms of the action sensitivity, selectivity, and fast fault isolation.

The distribution network is protected with the centralized protection program, in which there are centralized protection devices equipped at the 10 kV center station and several FTUs installed at the site for fault detection and isolation. Fig. 4.13 is the engineering application diagram, Fig. 4.14A shows the FTU, and Fig. 4.14B shows the centralized pilot protection. The FTU uses the IDP831 distribution network measurement and control terminal (with ONU built in), with IDP801 centralized pilot protection device for protection and EPON for communication, which features the single point to multipoint structure and is suitable for communication of overhead distribution automation without requirement for synchronization. Based on the EPON transmission GOOSE mechanism, FTU uploads the starting and direction information to the centralized pilot protection device via GOOSE, and the centralized device receives the overcurrent and direction state information sent by FTUs in case of a fault, positions the fault section in the regional distribution network, and isolates the fault. Compared with the traditional fault isolation program with current protection, the recloser and sectionalizer, this program can meet the requirement for connection of the new energy into the active distribution network, and allows for faster and more accurate fault isolation.

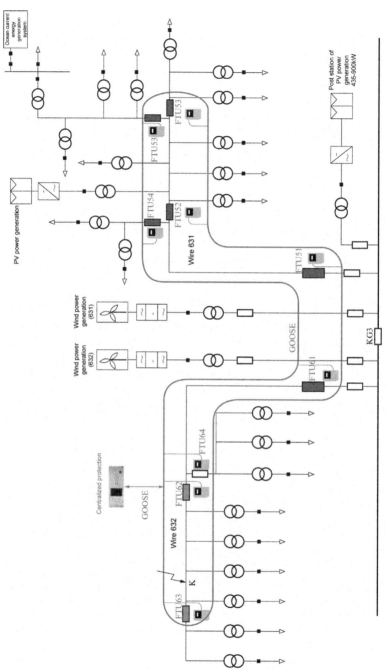

FIG. 4.10 Wiring of distribution network in Nanji Island microgrid demonstration project.

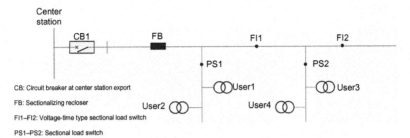

FIG. 4.11 Wiring of overhead single radiation line.

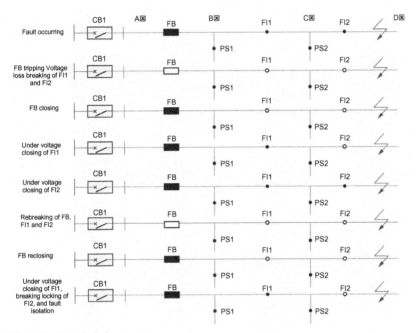

FIG. 4.12 Fault isolation process.

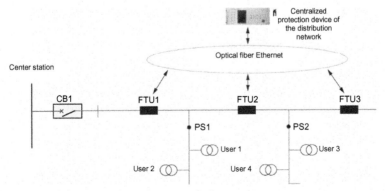

FIG. 4.13 Engineering application of centralized pilot protection.

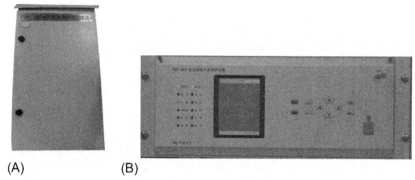

(A)   (B)

FIG. 4.14 FTU and centralized protection device of distribution network. (A) FTU and (B) centralized protection device.

4.2 Differential protection for cable line active distribution network

The overhead active distribution network protection program is introduced in Section 4.1. This section will introduce the active distribution network protection program for the cable line, since a different FA protection program of distribution network is adopted for different overhead wiring modes and the cable network wiring mode of the distribution network for DA.

4.2.1 Influences on cable distribution network caused by connection of distributed resources

The four common cable network connection modes: the double circuit radial system with one-side power supply, the one circuit system with two-side power supply, the double circuit system with two-side power supply, and the radial system with two-side power supply. The cable network is installed with the switch of the circuit breaker and the ring main unit. The feeder automation protection programs for the cable distribution network include two types of the distributed FA and the centralized FA of the cable network.

In the distributed FA program of cable distribution network, protection of the distribution network is realized with the feeder sectional program at site. The ring main unit is installed with distribution terminal units (DTU), which are connected by the special communication channel for point-to-point communication between the neighboring DTUs. For the sectional switch, the fault is identified through identification of the overcurrent state of the neighboring DTUs. The closing state will be kept if the overcurrent state of the neighboring DTU branches is consistent. Otherwise, the fault section will be identified, the DTU will trip and the fault will be isolated. Fig. 4.15 shows handling of the fault in the double power supply ring network. When K_1 fails, the overcurrent state of neighboring branch is inconsistent with that of DTU1-2 branch; the

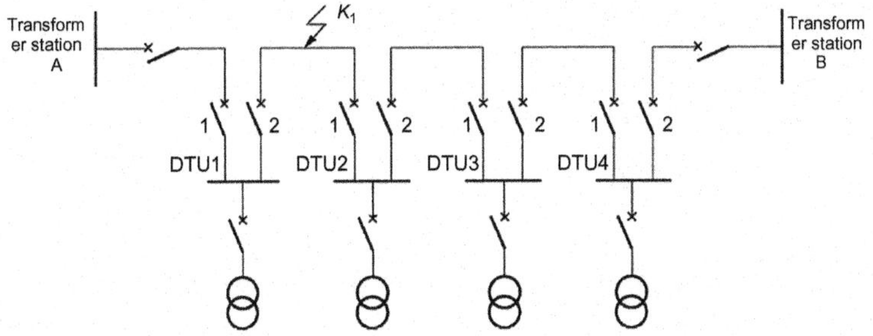

FIG. 4.15   Fault handing of double power supply ring network.

fault is detected between the branches of DTU1-2 and DTU2-1, DTU1-2, and DTU2-1 will trip; and the fault will be isolated. For the interconnection switch, the fault is identified through identification of the overcurrent state of the neighboring DTU. The sectional switch will be closed if the overcurrent state of the neighboring DTU branch is consistent. Otherwise, it will be kept opened.

This FA protection program allows for ideal fault isolation without distributed resources connected. It is impossible to identify the fault effectively with the distributed resources connected. For example, if the distributed resources are connected to DTU2 branch, when K_1 fails, the distributed resources will supply fault current which will flow through DTU2-1, and the overcurrent state of the neighboring DTU is consistent. Hence, it is impossible to identify the fault section for fault isolation.

In the centralized FA program, protection of the distribution network is realized with the help of the main station (substation) and the feeder terminal. Take Fig. 4.15 as an example again. The ring main unit is installed with DTU, which is connected with the main station (substation) via the communication system.

Without the distributed resources connected, in case of a fault inside the section, the outgoing circuit breaker of the substation trips due to the overcurrent, and the instantaneous fault reclosing is successful. If the permanent fault reclosing fails, fault handling of the main station (substation) will be started, DTU will transmit the fault information to the main station (substation), which will judge the overcurrent state of the neighboring branch. If the fault is identified within the section with the inconsistent overcurrent state of branch, the main station (substation) will break the corresponding DTU via remote control and isolate the fault section.

With the distributed resources connected, in case of a fault at the side of power supply outside the section, the distributed resources supply the fault current, which will flow through the outgoing circuit breaker, thus causing the incorrect action of the outgoing circuit breaker. In case of an instantaneous fault in the section, after the outgoing circuit breaker trips, it could cause failures of arc extinguishing and reclosing due to the existence of the distributed resources; in the case of a permanent fault, it will result in the failure of reclosing, the main station (substation) will start fault handling, and FTU will transmit the fault information to the main station (substation). Due to the existence of the distributed resources, the main station (substation) judges that the overcurrent state of the neighboring branch is consistent. Hence, it is impossible to identify the fault section for fault isolation.

4.2.2 Differential protection program for active distribution network

In the power transmission network, the bidirectional current is a very common power transmission mode. The relay protection technology of the power transmission network is an ideal solution. The differential protection is widely applied for protection of the lines, bus, transformers, and generators in the power transmission network. As the most simple and reliable protection mode, it is used as the main protection for the objects above and has been explained in many articles. Hence, it will not be repeated herein. Along with the development of the smart substation technology, the intelligent electric devices (IED) can realize synchronous sampling. The development of the network technologies and the computer technologies allows for the centralized protection of the regional network. Thus, the differential protection technology of power transmission network can be introduced for the differential protection of the active distribution network.

4.2.2.1 Smart substation synchronization technology of IEEE1588 (or B code)

The interlayer protection control devices and the process layer smart terminal devices of the smart substation are collectively referred to as *IED*. The IEEE1588 signal and IRIG-B code signal can meet the demand for microsecond precision for time synchronization of the interlayer and process layer of the smart substation.

IEEE1588 time synchronization adopts the distributed measurement method and the precision time protocol (PTP), to synchronize the clocks independently running at the measurement separation nodes to a clock with higher accuracy and precision via the network connection based on IEEE1588 standard, which can solve the problem of clock synchronization for the network. In the smart substation, the clock synchronizer receives the GPS signals, which will be transmitted to the network switch (supporting 1588 time synchronization) as the master clock source, connects the IED on the process bus as the slave clock, and synchronizes the clock via the process layer network based on 1588 protocol, so as to synchronize the IED of the smart substation to the unified clock source. The 1588 time synchronization network needs a multiport network switch for physical connection (see Fig. 4.16).

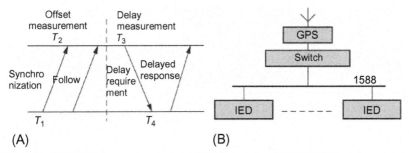

FIG. 4.16 IEEE1588 time synchronization. (A) PTP principle and (B) IEEE1588 process network.

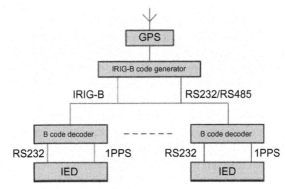

FIG. 4.17 IRIG-B code time synchronization.

IRIG-B code is the clock code specially developed for clock transmission by IRIG. In the smart substation, IRIG-B code time synchronization is a process based on B code generator, in which the RS232 data transmitted by the GPS receiver and the 1PPS output is converted into IRIG-B code for exporting from the IRIG-B code export and the RS232/RS422/RS485 serial interface. The IED for time synchronization converts the B code into the standard time information and 1PPS pulse signal via the B code decoder. IRIG-B time synchronization network needs a multiport B code generator for physical connection (see Fig. 4.17).

4.2.2.2 Multiend differential protection synchronization technology for the power transmission line

The current differential protection synchronization technology of multiend differential protection for the power transmission line is based on the current differential protection of the fiber channel, and realizes synchronous sampling for dual-end differential protection of the power transmission line. Besides the dual-end differential protection for the power transmission line, the three-

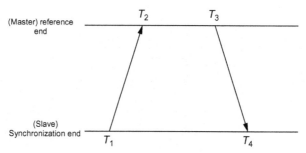

FIG. 4.18 Principle of channel-based synchronization.

end differential protection of the teed lines has also been widely used, for which the synchronous sampling can be realized via the fiber channel. In the new relay protection with the four-end double-circuit lines on the same pole consisting of the fiber channels, the four-end protection of the double-circuit lines on the same pole realizes synchronous sampling via the fiber channel ring connection, which can form the longitudinal differential protection and the transverse differential protection.

Fig. 4.18 shows the channel-based synchronization. The (slave) synchronization end sends the synchronization order at the time of sampling, and the (main) reference end receives the synchronization order at the time of T_2 and sends the reference information to the synchronization end at the sampling time of T_3. The synchronization end receives the reference information at the time of T_4 and calculates the channel delay and sampling deviation.

The channel delay is

$$T_{\text{Delay}} = ((T_2 - T_1) + (T_4 - T_3))/2 \qquad (4.7)$$

The sampling deviation is

$$T_{\text{Offset}} = ((T_2 - T_1) - (T_4 - T_3))/2 \qquad (4.8)$$

The synchronization end corrects the sampling time based on the sampling deviation for sampling synchronization.

4.2.2.3 DTU synchronization of active distribution network with built-in Ethernet switch

The smart substation synchronization technology based on the IEEE1588 (or B code) signal can realize synchronization of IED at the smart substation. This technology adopts the multiport network switch, and is suitable to the devices collectively installed inside the substation in view that IEDs in the substation are close to each other. The ring main unit is a "hand-in-hand" ring network structure with the devices far from each other. To adapt to the "hand-in-hand" ring network connection of the ring main unit, the DTUs are built with the Ethernet (i.e., each DTU serves as a network switch with two optical Ethernet interfaces). Each optical Ethernet interface has the function of transmitting

(TX) and receiving (RX), and can realize "the hand-in-hand" connection between the DTUs..

Fig. 4.19 shows the network mode of DTUs in the active distribution network with built-in Ethernet switch. This network structure is simple with dedicated fiber channel arranged for the cable lines. Fig. 4.20 shows the built-in Ethernet interface board.

There are two modes for main clock selection: in the first program, a switch supporting IEEE1588 is installed in the substation. The switch in the substation works as the master clock source, and the DTU on the outgoing line of the substation is connected with the time synchronization network switch in the substation, so as to realize the synchronous data collection at each node of the ring main unit. This program requires an IEEE1588 switch configured in the substation, with the switch in the substation as the master clock source and the switch in the DTU as the slave clock. In the second program, instead of the IEEE1588 switch, a DTU is selected randomly as the master end, the other DTUs serve as the slave ends, for realizing synchronization with the fiber channel.

4.2.2.4 Centralized line differential protection and decentralized bus differential protection

The current differential protection based on Kirchhoff's current law has been widely applied in high voltage line protection and bus protection. It is a simple

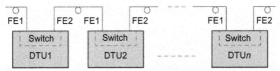

FIG. 4.19 "Hand-in-hand" networking mode of DTUs (FE1 and FE2 with built-in optical Ethernet interfaces).

FIG. 4.20 Built-in Ethernet interface board.

and ideal protection solution. The differential protection of power transmission line only requires synchronization of the current information at both sides of the line, and is realized via comparison of the current at both sides of the line. The bus protection is realized based on the differential region formed by the current flowing into and out from the bus. Hence, the centralized line differential protection and decentralized bus differential protection can be arranged based on the wiring characteristics of the ring main unit in the distribution network.

The differential protection region model is built according to the cable connection in the double power supply ring distribution network in Fig. 4.15, as shown in Fig. 4.21, in which DG is the distributed generation, LD refers to the load, Ln is the differential region of the distribution network line, and Bn is the bus differential region of the ring main unit. The synchronization sampling current data (e.g., the substation outgoing line current for synchronization sampling in the differential region of Ln line, the incoming and outgoing current of DTU in the ring main unit) are transmitted to the centralized differential protection device via the "hand-in-hand" concatenated special optical fibers, for realizing the centralized differential protection function of the distribution network lines.

The criteria for differential protection of Ln power distribution network lines are as follows:

The start criterion is

$$I_d > I_{QD} \tag{4.9}$$

The ratio restraint criterion is

$$I_d > k I_r \tag{4.10}$$

where I_d is the differential current, $I_d = |I_1 + I_2|$; $I_r = |I_1 - I_2|$ is the restraint current; k is the ratio of restraint; and I_{QD} is the threshold of differential current action. I_1 and I_2 are the current at sides of the ring network line. The restraint value, which is equal to the absolute value of the difference between the currents at sides of the ring network line, can effectively improve the fault sensitivity within the section.

For the local ring main unit, to reduce the volume of synchronization data for centralized differential protection, the bus differential protection program with

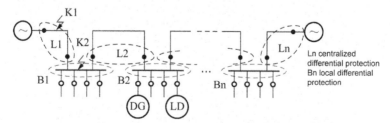

FIG. 4.21 Model of differential protection section of distribution network line.

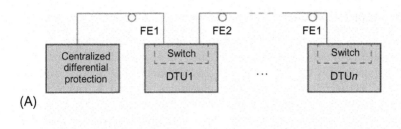

(A)

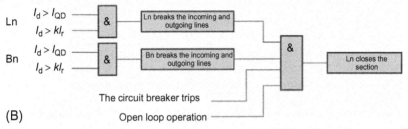

(B)

The circuit breaker trips

Open loop operation

FIG. 4.22 Logic diagram of device connection and protection. (A) "Hand-in-hand" connection of DTU and centralized protection device and (B) logic diagram of protection.

DTU on site is adopted for handling of the bus fault of the ring main unit. Bn is the differential protection section of the ring main unit. The criteria of differential protection are the same as the Formula (4.9) and Formula (4.10). $I_{\mathrm{d}} = |\sum_{j=1}^{n} I_j|$; $I_{\mathrm{r}} = \sum_{j=1}^{n} I_j$; I_j is the branch current on the bus.

Fig. 4.22A shows the physical connection between DTU terminal and the centralized differential protection device. Fig. 4.22B is the logic diagram of protection. In case of closed loop operation in the ring network line, if the distribution network line fails, the centralized differential protection device will identify the fault, notify the corresponding DTU to directly break the circuit breakers at sides of the fault line, when the bus of the ring main unit fails, the bus differential protection device at site of DTU will directly break the incoming and outgoing circuit breakers of the ring main unit, so as to isolate the fault as soon as possible. In case of open loop operation in the ring network line, if the distribution network line fails, the centralized differential protection device will identify the fault, notifying the corresponding DTU to directly break the circuit breakers at sides of the fault line. After the centralized differential protection device trips, the protection device will notify the corresponding DTU to directly close and open the circuit breaker, so as to isolate the fault and recover the power supply as soon as possible. If the bus of the ring main unit fails, the bus differential protection device on site will directly break the incoming and outgoing circuit breakers of the ring main unit, and the centralized differential protection device will notify the corresponding DTU to directly close and open the circuit breaker.

4.2.3 Application Cases

Fig. 4.23 is the wiring of active distribution network of the microgrid demonstration project of the smart grid research institute of State Grid Corporation of China, which adopts the 10 kV dual ring wiring mode. The microgrid part is composed of Section I, where low voltage bus (400 V) is connected to 300 kW/600kWh lithium iron phosphate for energy storage, 450 kW PV power generation, 720 kW load; and Section II, low voltage bus (400 V) is connected to 450 kW PV power generation system, 720 kW load. The two sections of the low voltage bus are connected by the bus tie switch. In this project, the IEEE1588 network clock is used for synchronization of DTU, and the active distribution network is realized with the centralized differential protection device and optical fiber communication. In the ring network power supply, after a fault occurs, the centralized differential protection device will identify the fault section, isolate the fault, and transmit the fault handling process information to the automatic master station system.

Fig. 4.24 is the wiring of Dongao Island microgrid demonstration project in the New Energy Microgrid Demonstration Project on Wanshan Island of

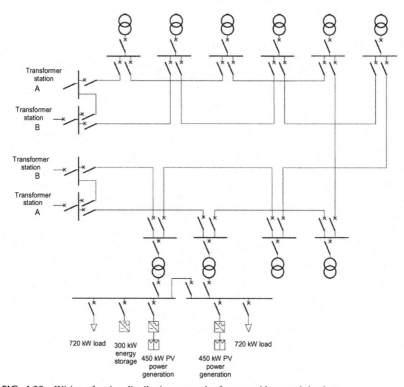

FIG. 4.23 Wiring of active distribution network of smart grid research institute.

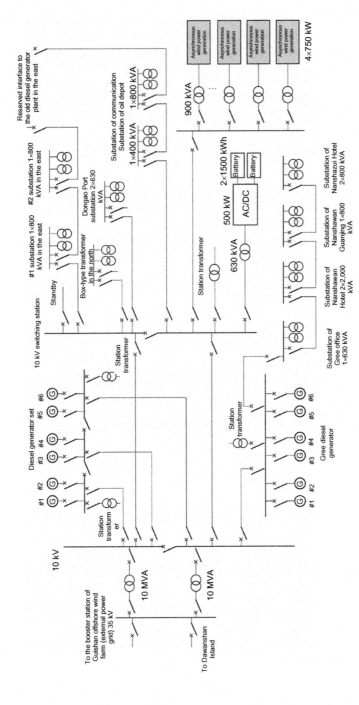

FIG. 4.24 Main wiring of microgrid demonstration project on Dongao island.

Zhuhai. The 10 kV cables are used for wiring in this project. The power of the diesel power generation is 6×1000 kW, that of wind power generation is 4×750 kW, and that of energy storage is 500 kW/3000 kWh. (The power of Phase 1 diesel power generation is 2×1000 kW; the microgrid runs independently.) This project adopts the optical fiber synchronization DTU, and there is a centralized protection device configured for the microgrid, so as to realize protection and control of the independent microgrid. Fig. 4.25 shows the DTUs and centralized protection devices used for the two projects provided.

In the two demonstration projects, the 10 kV distribution network is connected with cables, the centralized protection programs with DTU and the centralized protection devices configured are adopted for distribution network protection, The accurate fault positioning and isolation are realized via the differential protection. Instead of sampling data synchronization by the DTUs, the centralized protection system collects and controls the data of the distribution network with the optical fiber based on the differential synchronization technology. Fig. 4.26 is the diagram of engineering application.

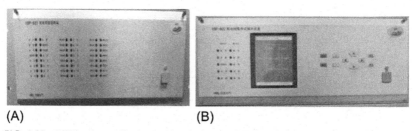

(A)   (B)

FIG. 4.25   DTU and centralized protection device. (A) DTU and (B) centralized protection device.

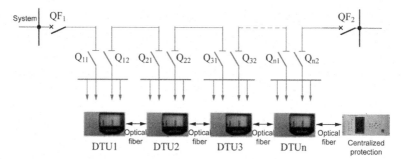

FIG. 4.26   Engineering application of centralized protection.

Chapter 5

Operation control and maintenance system of distributed resources

After connection of the distributed resources into the grid, it is contradictory to maximize the application of new energy and guarantee safe, stable, optimum, and economic operation of the distribution network. For the distribution network, the influences of the distributed resources on the distribution network will be minimized, and the distributed resources will be connected in an orderly way instead of the unordered way; for the distributed resources, the power generation efficiency of the new energy will be maximized for the change from restricted or fixed connection to the plug-and-play (PNP) connection. To solve this contradiction, it is necessary to establish an operation control system for the distributed resources, which can realize the orderly and scheduled power generation control and optimized dispatching of the distributed resources.

5.1 Operation control system of distributed resources

The capacity penetration (CP)—i.e., the ratio between the total capacity connected of the distributed resources and the total load of the system—is a key index currently used for evaluating the utilization rate of the distributed resources connected into the distribution network. However, the capacity penetration reflects the utilization degree of the distributed resources only from the view of the power generation capacity, without any consideration given to the time scale of DG connected into the distribution network for operation. If the penetration is very high but the effective operation time is short (i.e., the power generation is small), that the new energy resource will not be utilized effectively. Hence, it is more rational to evaluate the utilization degree of DG with "generating capacity penetration (GCP)," which is defined as the ratio of the power generation of the distribution resources and the total power consumption of the loads.

Distributed Power Resources. https://doi.org/10.1016/B978-0-12-817447-0.00005-5

This index evaluates the effective utilization degree of the distributed resources from two aspects of the capacity and the time. At present, CP is widely used for evaluating the utilization conditions of the distributed resources. However, it is difficult to give a specific standard index of CP due to the differences of the distribution network architectures. CP can be up to 100% in some distribution network, but can be lower than 10% in other ones. This is also the reason why there is no specific CP defined in the domestic or foreign standards. It is proper to pay more attention to the longer effective operation time of DG connected into the distribution network, thus generating more power.

The distributed resources feature wide distribution, remote location, demanding operation conditions, poor reliabilities of devices, heavy workload of O&M, and so on. The power generation quality and the economic efficiency will be influenced if the failure of a device has not been discovered in time. Besides, the theoretical analysis basis is insufficient for the operation efficiency of the distributed resources. Hence, it is impossible to realize the optimum operation. In the monitoring, control, operation, and maintenance program currently adopted for the distributed resources, the information is transmitted via the monitoring system, with due to the historical reasons, or by the dedicated interconnection interface device. The requirements for O&M and system configuration are inconsistent because of the large varieties of the monitoring and O&M system and the different connection system architectures, which makes it more difficult to control the distributed resources. Meanwhile, the demands for control and O&M have not been considered sufficiently.

To cope with the difficulties of decentralized operation, management and maintenance caused by connection of the distributed resources into the distribution network, a centralized management mode is introduced in this chapter for O&M of DG, in order to realize PNP of operation and maintenance. The centralized management is as follows: make a unified plan for O&M of the electrical devices of all the distributed resources at the sectional level, use the same monitoring platform, load all the dispersive panoramic data, monitor the operation conditions of all the devices on line, and deal with the fault alarm in a unified manner.

5.1.1 Collection of device terminal information

When the distributed resources are connected into the distribution network, the electric and electronic devices will work for energy conversion. All the electrical devices connected are named as the electrical device terminals, including (1) the power conversion devices for power supply switching (e.g., the inverter, the converter, and the charger); (2) safety interconnection devices, which are used for protection and safe isolation, and can meet the requirement of electrical interconnection, facilitate connection into the distribution network, and realize PNP connection into the distribution network; (3) the information interconnection devices, to transmit the information of devices in real time, realizing PNP control and O&M.

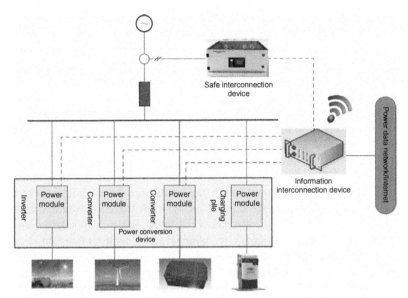

FIG. 5.1 Collection of device terminal information.

Fig. 5.1 shows the power conversion devices, the safe interconnection devices, and the information interconnection devices of the distributed resources. The information interconnection devices can not only realize communication collection of the distributed resources, but also allow for connection and information forwarding of BMS and a variety of external smart devices supporting IEC104, IEC61850, and Modbus protocols, and support data collection via the serial interfaces and Ethernet communication. The data is transmitted to the distributed resources control system and the O&M and control system for PNP information interconnection devices.

The information interconnection technology features "the distributed resource+Internet." Another function of the information interconnection devices is recognition of Win and QR codes. Setting adjustment and operation state monitoring of the device can be conducted by scanning the QR code of the connected electrical device with a mobile terminal. Commissioning and setting of the electrical device with a mobile terminal allows for contactless commissioning, O&M inspection and diagnosis, which can ensure the safety of personnel and devices, eliminate the hazards of climbing and electric shock during O&M of the pole-mounted test, and control terminals, thus maximizing the safety of the operation.

5.1.2 Operation control system of distributed resources

The operation control system of a distributed system can be used as a subsystem of the automation system for the distribution network. It can serve as a separate operation control system of the distribution resources, or be built into the automation system of the distribution network. It can realize the orderly connection and operation control of different distributed resources based on the function of different types of distributed resources connected. Its function includes the PV power generation control, the wind power generation control, energy storage control, and charger control. It can realize PNP connection of the distributed resources instead of the restricted or fixed connection and orderly control, or the unordered control. Fig. 5.2 shows the control communication architecture of the distributed resources system. The device terminal information collected from the information interconnection devices are transmitted based on the IEC 61850 protocol. The non-IEC 61850 protocol information will be converted into the IEC 61850 protocol information by the information interconnection device before being transmitted to the control system of distributed resources. In the absence of a specific type of distributed resources, the corresponding control function is canceled. For example, the wind power generation function is canceled if there is no wind power generation in the section under control. Fig. 5.3 shows the control of distributed generation with input of the forecast information, including the PV power generation forecast, the wind power generation forecast, and the load forecast. In this system, safe and economic operation of the distribution network is realized from the aspect of the distribution network, and the power generation efficiency is maximized from the aspect of connection of the distributed resources. Besides comprehensive and ordered control

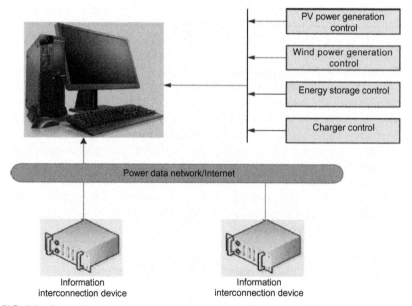

FIG. 5.2 Communication architecture for control of distributed resources system.

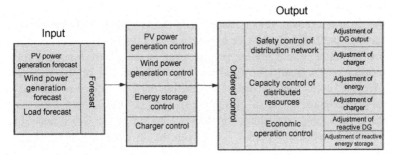

FIG. 5.3 Control of distributed resources.

is realized through the PV power generation control, wind power generation control energy storage control, and charger control.

The output of the distributed resource, the controllable load (charger), and energy storage is adjusted for safe and economic operation of the distribution network and improvement of the capacity for the distributed resources.

5.1.3 O&M system of distributed resources

To cope with the difficulties of decentralized operation, management, and maintenance caused by connection of the distributed resources into the distribution network, a centralized management mode is introduced for O&M of DG, in order to realize PNP function for operation and maintenance. The system architecture for O&M of the distributed resources is as shown in Fig. 5.4.

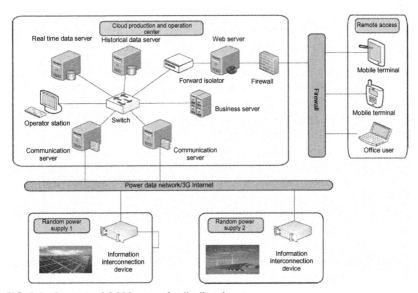

FIG. 5.4 Structure of O&M system for distributed resources.

The real-time data of the distributed resources is transmitted to the O&M system via the dedicated power data network. The O&M system is connected to the Internet via the firewall and can remotely access the operation conditions of the distributed resources via the mobile terminal. It mainly functions for (1) monitoring of the real-time data and state information of the distributed resources; (2) collection and analysis of the big data; (3) fault warning and historical alarm inquiry; and (4) designing of the O&M tasks and work orders based on the O&M procedure, for implementation of the O&M tasks in an orderly and acceptable manner. Along with the development of big data and cloud computing technologies, the O&M system based on the cloud architecture can be adopted for the distributed resources system.

The real-time data of the distributed resources needs to be collectively transmitted to the O&M system and the control system of the distributed resources at the same time, which will be set respectively based on their function. The control system of the distributed resources focuses on the distributed generation control within a specific distribution network section, and serves as a subsystem of the automation system of the distribution network. The cloud operation and maintenance center focuses on O&M of the distributed generation within multiple distribution network sections.

The control system of the distributed resources and the O&M system of the distributed resources are independent of each other without information transfer. In the actual engineering application, different arrangement modes (e.g., the control system of distributed resources + distributed resources, the O&M system of distributed resources + distributed resources, the control system of distributed resources + distributed resources, and the O&M system of distributed resources) can be adopted according to the specific requirements.

5.1.4 Operation control system of distributed resources (PV)

The operation control system of the distributed resources can be used separately or in a combined manner for different objects of distributed resources (e.g., the operation control system of distributed PV generation, the control system of distributed wind power generation, the control system of distributed energy storage, the control system of distributed charging, or the comprehensive control system of the distributed wind generation, PV generation, and energy storage). A case study on the control system of the distributed PV power generation will be briefly introduced as follows. The other systems are similar to the control system of the distributed PV power generation, which will not be repeated herein.

The control system of distributed PV power generation is suitable for the centralized control of distributed PV power generation, including the PV power generation monitoring, AGC for power dispatching control of PV power generation, AVC for reactive control of voltage, and PV power generation forecast, which can realize the connection control of the high CP distributed PV power generation.

5.1.4.1 PV power generation monitoring

The PV power generation monitoring system monitors the PV power generation devices, the grid-connected inverters, and the auxiliary AC/DC devices in the PV power generation system, and analyzes the conditions of grid for connection, the power generation quality, and the power generation volume. This can realize the operation monitoring and power generation control of the PV power generation system.

The PV power generation monitoring system monitors the operation conditions of the devices in real time, and triggers an audible and visual alarm on the monitoring interface for the abnormal signal generated by the device. This monitoring system can also monitor the power quality indexes in real time, including the voltage, frequency, power factor, three phase unbalance rate, the harmonic content and other critical ones, so as to make sure the power generation system supplies electric energy of reliable quality. Monitoring of PV power generation includes three parts of SCADA of PV power generation, power generation statistics, and generated energy management.

The main functions of SCADA of PV power generation include data collection and processing, event and alarm, PV inverter operation monitoring, operation monitoring of the PV box type transformer, operation monitoring of the PV junction box, and power quality monitoring.

Statistics and analysis of PV power generation include the current power generation, daily power generation, monthly power generation, accumulated power generation, accumulated CO_2 emission reduction, and generated power statistics and analysis.

Generated energy management includes locally and remotely controlled energy management. The locally controlled energy management includes the exchange power curve control, the smooth control active output, the reactive automatic voltage control, the emergency supporting dispatching of power, and the emergency shutdown dispatching; and the remotely controlled energy management includes the exchange power curve control, the emergency supporting dispatching of power, the emergency shutdown dispatching, and the reactive automatic dispatching of voltage. Fig. 5.5 shows an interface of PV power generation monitoring. Table 5.1 lists the function setting of the PV monitoring system. The main control function of power generation management is as follows.

(1) Exchange power curve control

The local exchange power curve control: In the grid-connected operation, the PV power generation output is controlled properly based on the planned value preset of the exchange power curve, thus realizing the operation with the specified exchange power under the premise of ensuring economic and safe internal operation of the grid.

The remote active power dispatching control: The local generation power is adjusted based on the remotely dispatched power setting or power curve within the specified time. A time limit is defined for the remotely

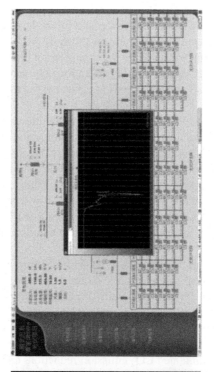

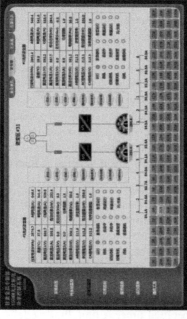

FIG. 5.5 An interface of PV generation monitoring.

TABLE 5.1 Function setting of PV generation monitoring system

Function type	Function name	Function type	Function name
SCADA	Data collection and handling	Energy management	Local control—exchange power curve control
	Database establishment and maintenance		Local control—smooth control of active power output
	Control operation		Local control—reactive automatic control of voltage
	Alarm processing		Local control—emergency supporting dispatching of power
	Interface generation and display		Local control—emergency shutdown dispatching
	Online calculation and tabulation		Remote exchange power curve control
	Self-diagnosis and self-recovery of system		Remote control—emergency supporting dispatching of power
Power generation statistics	Monitoring, statistics, and analysis of PV power generation		Remote dispatching—emergency shutdown dispatching
			Remote dispatching—reactive automatic dispatching of voltage

dispatched setting and curve. When the time limit expires, the local power will automatically recover to the normal operation state. The remote active power dispatching controls the issued setting and the curve, and determines the plan type as required.

(2) Emergency power supporting control

The local emergency power supporting: The local dispatching mode is adopted. When the system is in urgent need of a large power to support the grid, the emergency power supporting control can be selected for PV generation. After receiving the order, the PV generation system will maximize the power generation and minimize the power consumption as soon as possible, so as to meet the dispatching requirement.

The remote dispatching emergency supporting control: The remote dispatching mode is adopted. When the remote dispatching is in urgent need of a large power to support the grid, the system can perform emergency supporting control of power for PV generation. After receiving the order, the PV generation system will maximize the power generation and minimize the power consumption as soon as possible, so as to meet the dispatching requirement.

(3) Emergency generation shutdown control

The local emergency shutdown control: The local dispatching mode is adopted to shut down all the generation devices as soon as possible, so as to meet the dispatching requirement.

The remote emergency generation shutdown: After receiving the remote emergency shutdown order, the PV generation system will stop all the PV generation devices as soon as possible, so as to meet the dispatching requirement.

5.1.4.2 PV power generation forecast

The PV power generation forecast system can analyze and forecast the active power of PV power generation in advance for a certain period by virtue of the statistical laws and other technologies and methods with the historical power generation data, the numerical weather forecast data, the real-time meteorological data collection, and the operation conditions of the PV power generation devices based on the meteorological forecast data, and build the forecast model of PV power generation for forecasting of the ultra-short-term power generation power in the coming 15 min~4 h and that in the coming 72 h.

The PV power generation is influenced by the meteorological factors of the weather conditions, seasonal variation, solar radiation level, cloud coverage, temperature, and so on. The PV forecast algorithm is based on the historical meteorological data (solar radiation level, temperature, and others) and the PV power generation data during the same period in the past years. It is a power statistics forecast method with an analysis model built with the statistical method (e.g., the neural network algorithm and the vector machine algorithm) and the numerical forecast result entered. The forecast precision depends on the crucial meteorological factors and influence the PV power generation, the correlation analysis, and the precision of meteorological forecast. The forecast structure of PV power generation is as shown in Fig. 5.6. The forecast is made with the short-term forecast model and the ultra-short-term forecast model based on the numerical weather forecast and the local meteorological information entered, with the results of ultra-short-term power forecast and the short-term power forecast output.

(1) Ultra-short-term power forecast

The time scale of ultra-short-term power forecast is 0–4 h. Basically, the movement of cloud layers is forecast according to the satellite cloud

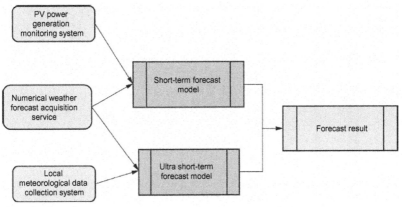

FIG. 5.6   Forecast structure of PV power generation.

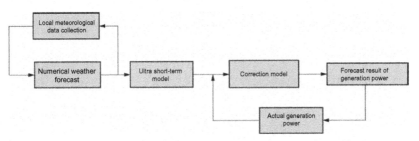

FIG. 5.7   Mathematical model for ultra-short-term forecast of PV power generation.

images taken by the geosynchronous satellite, which is given for forecasting of the cloud layer indexes in the coming hours.

Then, the forecast ground irradiation intensity can be obtained by virtue of the linear relation between the cloud layer index and the ground irradiation intensity, and the forecast output power of PV power generation can be worked out with the efficiency model.

Fig. 5.7 is the forecast power model of the ultra-short-term PV power generation. The directly obtained numerical weather forecast (mainly including the solar radiation level and the temperature) is used as the input of the neural network or vector machine algorithm for forecasting of the generator power in the coming 0–4 h. A regression analysis is made with the model for the forecast power and the actual power, for correction of the next ultra-short-term forecast result. Fig. 5.8 shows the result of predicted power of the ultra-short-term PV power generation in an actual PV power generation system at site of the SOFT—8000 PV power generation system, where Curve 1 represents the predicted value and Curve 2 represents the actual value. The generation power is forecast every

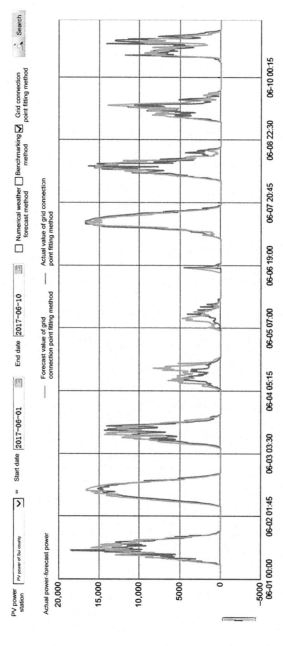

FIG. 5.8 Ultra-short-term result of PV power generation forecast.

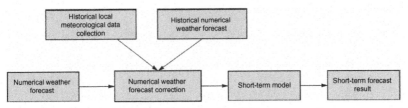

FIG. 5.9   Mathematical model for short-term forecast of PV power generation.

15 min, to realize a forecast precision above 90% for the ultra-short-term power generation finally.

(2) Short-term power forecast

The time scale of short-term power forecast is 0–72 h. In general, the numerical weather forecast (mainly including the solar radiation level and the temperature) of the coming 1–3 days needs to be obtained according to the medium-scale numerical weather forecast. Then, the forecast ground irradiation intensity is obtained based on the historical data and the meteorological element information, to work out the forecast output power of the PV power generation.

Fig. 5.9 is the forecast power model of the short-term PV power generation. The numerical weather forecast is corrected using the directly obtained numerical weather forecast of the coming 1–3 days (mainly including the solar radiation level and the temperature), the historical local meteorological data is collected, along with the historical numerical weather forecast. The short-term PV power generation output power is forecast with the short-term forecast model. Fig. 5.10 shows the forecast power result of the actual short-term PV power generation at site, wherein Curve 1 represents the forecast value and Curve 2 represents the actual value. The generation power of the coming 3 days is forecast. The short-term generation forecast precision is larger than 85% (the result in the diagram is above 93%).

5.2   Application of IEC 61850 in distributed resources

The distributed resources feature the flexible power generation mode, independence of the operation conditions of the grid, large consumption of the clean energy, and is consistent with the concepts of energy saving, emission reduction, green energy resources, and sustainable development. However, many challenges occur in operation of the grid connected with lots of the distributed resources. In order to ensure the operation stability, safety, and reliability of the grid connected with the distributed resources, it is necessary to define the unified communication and control interfaces for the devices and system of the distributed resources, so as to reduce the installation cost, simplify the deployment

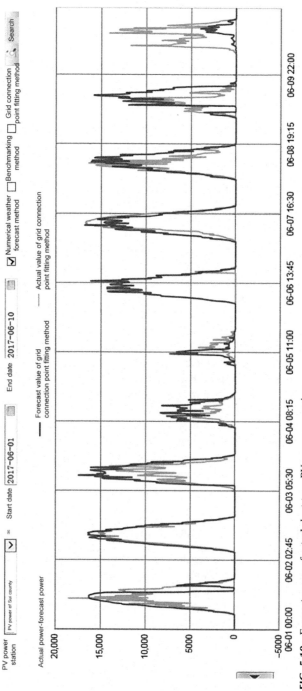

FIG. 5.10 Forecast power of actual short-term PV power generation.

implementation and O&M of the distributed resources, and improve the operation reliability of the grid.

IEC 61850 Edl.0, *Communication Networks and Systems in Substations*, was issued in 2004 and was formulated by Technical Committee 57 (TC57) of the International Electro technical Commission (IEC). It aims to utilize the information technology (IT) in the power system, in order to provide a communication standard for the automation system of the substation, realize the seamless connection between devices, PNP, and the ultimate goal of industrial control communication "one world, one technology, and one standard."

The name of IEC 61850 Ed2.0 was changed to *Communication Networks and Systems of Power Systems*. That is to say, its application scope is no longer restricted to the automation application field of the substation. An abundant standard system is formed, in which, IEC 61850-7-420 is a communication standard for the distributed resources systems, and is suitable for information exchange between the device terminal and the control center, model transformation, system integration, and conformity test. IEC 61850 is a critical supporting technology for construction and development of distributed generation, and is utilized to realize interconnection, intercommunication, and interoperation between the distributed generation and the management system, and PNP of the devices connected.

5.2.1 Introduction of IEC 61850

To cope with the failure of interoperation between IEDs of different manufacturers on the substation automation system (SAS), IECTC57 formulated the IEC 61850 standard. The first edition was released in 2004 and named as IEC 61850 Edl.0. This standard summarizes the experiences on product development and application of the substation automation system in the recent 20 years in the world, This standard also absorbs the latest IT technology achievements, and adopts lots of universal technologies of object-oriented modeling, upper-level communication services independent of the bottom level communication protocol.

After being released for 10 years, IEC 61850 has effectively solved the problem of interoperation and protocol conversion of the products for the substation automation systems, as well as allowing for the self-description and self-diagnosis of the substation automation devices, facilitating system integration and reducing the engineering cost of the substation automation system.

IECTC57 has cooperated with the standardization organizations to amend the existing international standard, draft the technical reports and technical codes, and continuously promote development of IEC 61850 technology, in order to meet the application demands of the smart grid. The relevant documents of IEC 61850Ed2.0 are released in the forms of international standards (IS) and technical reports (TR). For example, IEC 61850-7-420 is a standard on communication systems of the distributed resources, and IEC 61850-9-7 is a

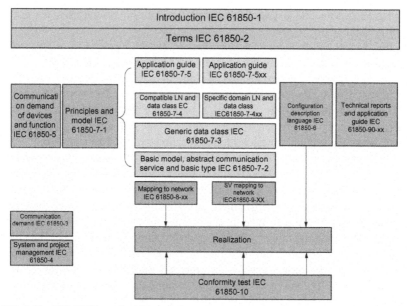

FIG. 5.11   IEC 61850 standard framework.

technical document on application of IEC 61850 object model of the inverter. Fig. 5.11 is an IEC 61850 standard framework.

5.2.2   Distributed resources and IEC 61850

There are a variety of devices connected to the distributed resources. With the traditional communication protocol, the information point tables need to be configured and checked manually; Since the workload of installation, commissioning, and maintenance is large; and it is urgent to solve the PNP problem. PNP means the distributed resources can be discovered by the main station (distributed resources management system) automatically after connection. The main station automatically receives the data model configuration information uploaded, and automatically completes the configuration information of the device in the main station database according to the configuration information.

Specialists and scholars have put forward the concepts of smart distribution network and active distribution network to cope with the connection of lots of distribution generation devices and the new requirements for operation of the distribution network. In essence, the secondary information system and the primary distribution system are highly integrated to realize the informatization, automation, interaction, and intellectualization of the distribution network. Realization of informatization is based on the information model and information exchange method. The existing power system communication protocol organizes the data with the information point table, without description of the data source and other data relation. Hence, it is impossible to realize interconnection, intercommunication, interoperation, and PNP between the devices and the systems, and the workload of installation and commissioning is large.

It is crucial to realize interconnection, intercommunication, and interoperation of the distributed generation devices and the automation system, and PNP of the distributed resources with IEC 61850, in order to promote construction of the distributed resources. IEC 61850 standard is the dominant communication protocol in the power sector at present, which can effectively integrate the devices of different types into the system. Besides, IEC 61850 defines the information model and information exchange method of IED, and can realize communication between the IED and the main station of system, and between the IEDs. The open communication mode can facilitate operation management of the distributed generation, to improve the reliable operation control level of the distribution network.

The IEC 61850-7-420 standard focuses on the monitoring demands of the distributed resources, and covers management of the distributed resources, the unit controller, the generation system, battery monitoring, grid connection, and so on. It supports monitoring of combined heat and power generation, PV power generation, energy storage, and other distributed resources, and plays an important role of guiding for design, testing, communication of distributed resources, and control of the interfaces. At present, the application of the IEC 61850-7-420 standard in the distributed resources system mainly focuses on model building, model mapping, and merging.

IEC 61850-90-7 is the technical guide for the application of IEC 61850 in the distributed resources system. It introduces the distributed resources systems of the PV power generation system, energy storage, and so on, and analyzes the operation and control modes of the distributed resources systems based on inverters. Furthermore, it deals with modeling according to IEC 61850 for the distributed resources system with the operation mode and control mode as the demands. It also analyzes modeling according to IEC 61850 for emergency control, reactive voltage control, frequency control, voltage management, and other business types. The logic nodes used are provided in each item, and those defined in IEC 61850-7-4 and IEC 61850-7-420 are expanded and supplemented.

The technical report IEC 61850-90-15, *Integration of Distributed Resources based on IEC 61850*, provides a general information model for the interaction between DER and the grid operator. Fig. 5.12 shows DER integrated with a five-layer device architecture, including the process layer, the field layer, the station control layer, the operation layer, and the enterprise layer in the upward direction. The overall infrastructure for connecting DER into the automation system of the distribution network is formed by introducing a DER unit, DER unit controller, DER system, and enterprise DER management system, based on this five-layer infrastructure.

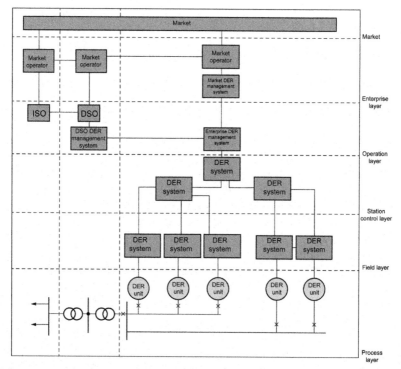

FIG. 5.12   DER integration infrastructure introduced in IEC 61850-90-15.

5.2.3   Plug-and-play

PNP aims to shorten the turning on time of the distributed resources, simplify engineering implementation and O&M, reduce the labor cost and workload, and meet the requirements for communication and control. The response and recognition mechanism between the terminal IED and the main station, automatic registration of IED, and automatic discovering of IED by the system are the preconditions for realization of PNP function.

The automatic discovering mechanism includes two types of automatic registration and automatic discovering, as shown in Fig. 5.13.

5.2.3.1   Automatic registration

After being put into operation, the new IED will send the registration information to the main station actively. After receiving the registration information of IED, the main station will inquire about the communication and configuration, communication address, and other relevant information of IED. If the terminal has completed the configured IED description (CID) document, it will call the corresponding CID document of IED, and configure IED information of the main station database according to the uploaded CID document. If the terminal

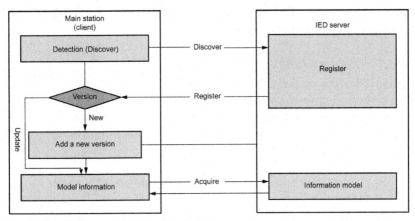

FIG. 5.13 IED discovering/registration mechanism.

has not configured the IED document, the main station will download the CID document to IED, which will organize the measurement and control information according to the CID document. After IED configuration information is updated at the site, the main station will be informed with the register mechanism.

5.2.3.2 Automatic discovering

After the main station sends the discover order, the newly connected IED will send the register information after receiving the discover order, and send or receive a CID document to/from the main station based on the configuration of the CID document. For the IED already installed, the configuration version information will be sent to the main station, and the main station will judge whether the CID document of IED has been updated based on the configuration version information received. If it has been updated, the updated CID document will be called.

5.2.4 Application challenges of IEC 61850

It is beneficial for the parties to apply IEC 61850 in the distributed generation. A series of critical technical problems needs to be solved, although it is consistent with the technical development direction.

5.2.4.1 Infrastructure

IEC 61850 introduces the three-layer device infrastructure of the substation automation system from the aspects of network communication and device function (i.e., the station control layer device, the interlayer device, and the process layer device). Hence, it is improper to indiscriminately copy the infrastructure mode for the distributed generation, and a suitable infrastructure will be designed based on the actual conditions. Efforts are being taken for this design both at home and abroad at present.

5.2.4.2 PNP function of terminal devices

The devices and system related will support PNP function in order to simplify the engineering implementation and maintenance of the distributed generation. It is necessary to design a reasonable response and identification mechanism to support the PNP function of the terminal devices according to ACSI supplement of IEC 61850 and a variety of data models.

5.2.4.3 Information exchange and coordination with main station system

At present, all the control center data models of the electric power systems in China are built according to IEC 61970/IEC 61968. A model built based on IEC 61850 is needed for the terminal devices of the distributed generation. Therefore, conversion and coordination with the model of the control center built based on IEC 61970/IEC 61968 and information integration with the management system are required.

5.2.4.4 Process of system integration

The document formats and templates defined in IEC 61850-6 Ed2.0 may not be directly used in the distributed generation system, since they are designed based on the features of the substation automation system.

5.3 O&M program of distributed resources based on the Internet of Things

5.3.1 Analysis for O&M of distributed resources

The O&M level of the distributed resources not only influences the long-term stable operation of the devices, but also relates to the operation cost, investment value, and the final income. The investors attach importance to the return on investment, and wish to minimize the time of operation in fault, so as to recover the cost as soon as possible. The grid operators focus on the influences of distributed resources on the grid, and have formulated standards to control and manage the connection of the distributed resources and improve the friendliness of devices connected into the grid. The users pay more attention to the application safety and reliability of distributed resources. Therefore, the O&M level of the distributed resources is significant to development of the new energy. Along with the development of the new energy, more and more devices of the distributed resources are connected into the grid, and they are distributed more broadly. The O&M mode based on scheduled or unscheduled manual inspection is time-consuming and labor-intensive, and features delayed repair

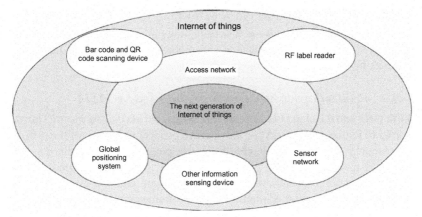

FIG. 5.14   Model of the Internet of Things.

and poor O&M level. Hence, it is insufficient to effectively improve the utilization level of the distributed resources.

The Internet of Things is a huge smart network consisting of the global positioning system, the sensor network, the barcode and QR code device, the radio frequency label reader, and other information-sensing devices, which are connected via a variety of access network and the Internet based on the agreed protocols. It is an information network for interconnection between things, and between people and the things. It can provide networked application and services with the smart interaction of device terminals as the core. Fig. 5.14 is a model of the Internet of Things. The Internet of Things can manage and control the personnel, machines, devices, and infrastructures in the whole network in real time, in order to manage production and life in a more refined and dynamic mode, bring higher levels of intelligence to the whole system, and improve the utilization rate of the resources and the productivity.

The O&M of distributed resources involves tracing of operation, maintenance (repair), and shutdown. The technology of the Internet of Things is based on the names defined as required. The devices information and operation conditions are uploaded to the cloud O&M center, where the device identity is recognized and positioned and repair is arranged, thus realizing life-cycle management for the devices.

5.3.1.1   Automatic recognition of devices

Automatic recognition technology is one of the technologies involved with the Internet of Things. Voice recognition technology, barcode recognition technology, and radio frequency identification technology included in the automatic recognition technology allows for setting the unique digital codes or differentiable identifications for devices of the distributed resources, which can facilitate rapid and valid recognition of the devices during O&M.

5.3.1.2 Accurate positioning of devices

The positioning technology is another technology involved with the Internet of Things. It allows for rapid and accurate positioning of devices of the distributed resources by virtue of the satellites, radio wave, and sensors.

5.3.1.3 Realization of PNP function and massive O&M

The devices and the cloud O&M center are connected via the Internet of Things, in order to realize smart O&M of the massive connected devices of the distributed resources and PNP function of the devices put into operation; this can also reduce the quantity of O&M personnel; improve the O&M efficiency of the devices; meet the demands of the investors, users, and the grid operators; and promote construction of the distributed resources.

5.3.1.4 Realization of life-cycle management

The Internet of Things technology is applied for life-cycle management of the distributed resources, in order to improve the utilization rate of devices and realize automatic and smart life-cycle management for the devices of the distributed resources.

5.3.2 PNP technology

In the Internet of Things, each device has a unique address (IP address), which is the identity in the digital world. The uniqueness and uniformity of the identity characteristics and codes is of great significance in operation of the Internet of Things. The unique identifications of devices can be identified with the images, voices, barcodes, radio frequencies, magnetic identification, biological characteristics, and so on. Sometimes, the identification methods can be used together

Information of connected device of distributed resources:

Device code (unique): XJ-20150520-001

Device name:

Device ——×××

Manufacturer: Xuji Group Corporation

Date of manufacture: March 5, 2015

QR code of device information:

FIG. 5.15 Unique identifier of device.

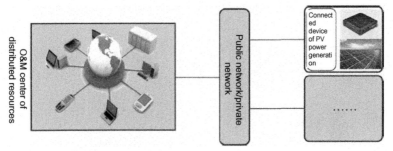

FIG. 5.16 Cloud O&M center of distributed resources.

for meeting the requirement of the actual application. In view of the advantage of low identification cost of the QR codes, a QR code can be set for a device as its unique identifier by the automatic identification technology. Fig. 5.15 is a device identifier of QR codes.

"PNP function" of a device means the cloud O&M center can automatically identify and manage the new device connected after it is put into operation. Fig. 5.16 is the cloud O&M center of the distributed resources. It is mainly manifested as the timeliness of application, and involves automatic and dynamic identification of the new devices, including automatic identification during initial installation, identification changed during operation, and automatic identification ended. For the cloud O&M center, PNP function for O&M of distributed resources refers to automatic identification and dynamic monitoring at the time the device is put into operation, without the complicated physical configuration and advance notice. After the distributed resources are put into operation, the cloud O&M center can automatically identify and manage the new device connected without manual configuration by the user. Since the code of the device connected is unique (simulated as a cellphone number), automatic identification and reporting of the position information and operation conditions of the device can be performed via active connection with the cloud O&M center after the device is put into operation. This function is similar to the case where a cellphone can be used only via a card insertion.

In view that the code of the connected device of the distributed resources is unique, after the device is put into operation, it will be actively connected to the cloud O&M center for identity recognition and confirmation. The operation conditions and the positioning information will be transmitted for management when the network connection is normal, thus realizing PNP function for O&M of the device connected (see Fig. 5.17). The geographic information system (GIS) is an interdiscipline that combines the disciplines of geography, cartography, remote sensing, and computers. It covers mapping and analysis of the phenomena and events on the earth, and integrates the map, which is a unique visual effect, and the geographic analysis function with the common database operations. The cloud O&M center integrates the position information transmitted by the distributed resources (the longitude and latitude information, and the

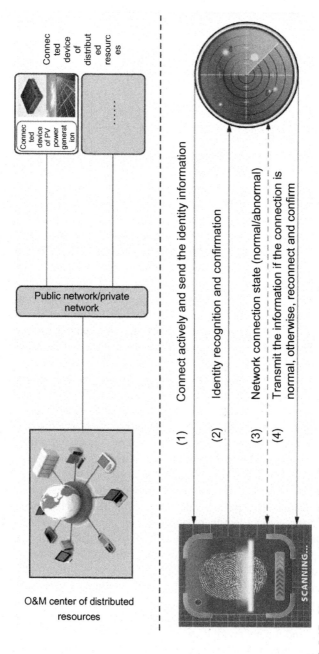

FIG. 5.17 Identity recognition of distributed resources.

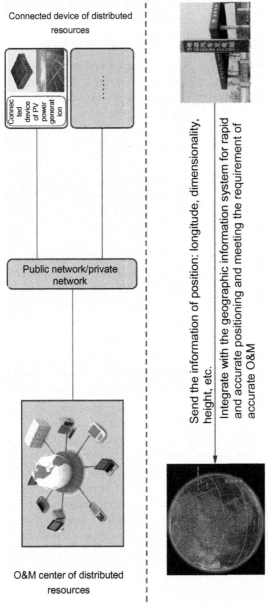

FIG. 5.18 Principle for positioning of connected device of distributed resources.

height information of GPS and Beidou positioning device) with GIS, for demonstrating the actual geographic scene of the device, realizing rapid and accurate positioning of the device in the geographic space and PNP function of device positioning in the real environment, as shown in Fig. 5.18.

5.3.3 Warning assessment of distributed resources

The O&M method in use at present is based on the previous experiences, in which scheduled or unscheduled manual inspection is performed for the system by collecting the types or O&M records and the handling methods, so as to guarantee device safety and system reliability. However, along with the extension and decentralization of the distributed resources, the shortcomings of such an O&M method become increasingly obvious (e.g., insufficient inspection; neglected points; untimely servicing; low efficiency, negligence, or error of inspection result recorded by hand; difficulties of learning the system conditions in a timely, accurate, and comprehensive manner; and formulating the best servicing and repair programs, which also impede further development of the distributed resources). The warning assessment method of the distributed resources combines the technologies for health condition assessment and the Internet of Things. The device condition information and repair records are sent to the cloud O&M center via the Internet of Things. The O&M expert system assesses the health conditions of the devices based on the health assessment model. If the health factor of the device is below the lower limit, a corresponding warning will be provided.

5.3.3.1 Health assessment model

The health assessment model of the distributed resources has four dimensionalities, including the service life of device, device fault (fault type, times of repair, and repair personnel), device warning (warning type and times), and use frequency (or normal operation time). The service life of the device is the benchmark, and the other three dimensionalities are used for fusion analysis, in order to work out the probability of one or several faults that may have recently occurred, and for weighing out the expected life so as to obtain the health assessment report of the device. The health assessment report mainly includes three aspects: the expected life, the correlation between the use frequency and the fault, and that between the use frequency and the warning. Fig. 5.19 is a multidimensional warning assessment model for the connected device of distributed resources.

5.3.3.2 Expected life of device

The expected life of device (L_{exp}) is the datum reference of the health assessment model. It is directly related to the service life (L_{age}), using frequency (calculated with different time scales, t, referred to as $F_{freq}(t)$) and the fault repair conditions (calculated with different time scales, the repair frequency is named

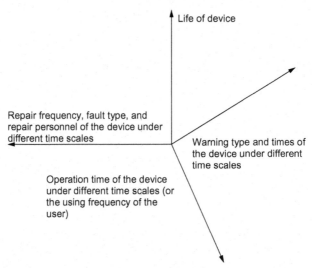

FIG. 5.19 Multidimensional warning assessment model for connected device of distributed resources.

$F_{rec}(t)$). The designed service life is named as L_{desi}; and the expected service life of the device is

$$L_{exp}(t) = \frac{L_{desi} - L_{age}}{1 + F_{freq}(t) + F_{rec}(t)} \qquad (5.1)$$

5.3.3.3 Analysis for correlation between warning and fault

The warming information, including the fault, warning, and other information about the device, will be transmitted according to the operation conditions after the distributed resources are put into operation. The general warning information is transmitted during normal operation, which will not influence the normal use of the device. The health assessment model can identify the type of warning, which is the most likely to cause a fault to the device, through analysis of the relation between the type of warning transmitted A_{type} (the occurrence frequency is calculated with different time scales $R_{freq}(t)$) and the type of fault repaired (the repair frequency of fault type $M_{freq}(t)$). For example, it is discovered from the warning history that the frequency of fault type T becomes very high when the warning type A is transmitted twice. Hence, it can be deduced that the correlation between the warning type A_{type} and the fault type T_{type} is very high. The correlation intimacy coefficient is expressed as C_{rel} (A_{type}, T_{type}). The accuracy of advance warning for repairing based on the device conditions can be improved via analysis, to avoid the unnecessary or excessive repair. After the warning type A_{type} within a specific time interval is transmitted, the probability P_{occur} (T_{type}, A_{type}) of fault T_{type} is

$$C_{\text{rel}}\left(A_{\text{type}}, T_{\text{type}}\right)(t) = \frac{A_{\text{type}}\left[R_{\text{freq}}(t)\right]}{T_{\text{type}}\left[M_{\text{freq}}(t)\right]} \tag{5.2}$$

$$P_{\text{occur}}\left(T_{\text{type}}, A_{\text{type}}\right)(t) = A_{\text{type}}\left[R_{\text{freq}}(t)\right] * C_{\text{rel}}\left(A_{\text{type}}, T_{\text{type}}\right) \tag{5.3}$$

5.3.3.4 Analysis for correlation between using frequency and device fault

It is necessary to analyze the correlation between the frequency U_{freq} and the fault type T_{type} of the device, to find out the nodes of the device liable to faults, obtain the relation between the fault type and using frequency, improve the accuracy of repair warning, shorten the repair time, and put forward the repair health warning based on the frequency of the device. The correlation intimacy coefficient of the fault type and the using frequency is expressed as $C_{\text{rel}}(T_{\text{type}}, U_{\text{freq}})$. Note that this coefficient is obtained according to analysis of the historical records. It is in proportion to the repair frequency of the device caused by the fault, and is inversely proportional to the using frequency within the time interval. It is a relatively constant value (e.g., $C_{\text{rel}}(T_{\text{type}}, U_{\text{freq}})(t_{\text{h}})$), worked out based on the historical time t_{h}. The correlation between the probability $P_{\text{occur}}(T_{\text{type}}, U_{\text{freq}})$ of fault type T_{type} within a specific time interval and the using frequency of the device is

$$C_{\text{rel}}\left(T_{\text{type}}, U_{\text{freq}}\right)(t_{\text{h}}) = \frac{T_{\text{type}}\left[M_{\text{freq}}(t_{\text{h}})\right]}{U_{\text{freq}}(t_{\text{h}})} \tag{5.4}$$

$$P_{\text{occur}}\left(T_{\text{type}}, U_{\text{freq}}\right)(t) = U_{\text{freq}}(t) \cdot C_{\text{rel}}\left(T_{\text{type}}, U_{\text{freq}}(t)\right) \tag{5.5}$$

5.3.3.5 Analysis for correlation between using frequency and device warning

The relation between the using frequency and device warning is analyzed to find out the habits most commonly leading to incorrect operation of the device, prevent the warning of incorrect operation via amendment of the operation instructions or improvement of the device, and adjust the order of warnings as per the use frequency of the device. The relation between the level L_A_{type} of the warning type within a specific time interval and the using frequency U_{freq} ($L\_A_{\text{type}}\_N$ is the original warning level) is

$$L A_{\text{type}}(t) = \frac{L A_{\text{type}} N}{U_{\text{freq}}(t)} \tag{5.6}$$

5.3.3.6   Repair warning expert system of distributed resources device

The repair warning expert system analyzes the health condition of the device based on the historical repair information and operation conditions of the device. The service life and device fault is the biggest concern of the repair personnel. If the expected life of the device is zero or a negative value, it indicates that the device needs to be scrapped, and will be replaced or dismantled as soon as possible. The probability of device fault mainly relates to the service life, which uses frequency and repair frequency of the device. If the probability of device fault obtained via analysis is larger than 80%, the expert system will send the geographic position of the device and the type of the expected fault to the O&M personnel for advance repair.

5.3.4   Procedure for O&M of distributed resources

The electronic O&M procedure can realize life-cycle management for the device, decrease the quantity of O&M personnel, and meet the demands of the investors, the users, and the grid operators. The warning expert system can provide device health assessment periodically with the multidimensional health assessment model, and send the assessment result to the cloud O&M center. If the assessment result is smaller than the eligible value, the cloud O&M center will send the type and probability of the possible fault and the geographic position of the device to the O&M personnel, so as to repair the device in advance (see Fig. 5.20).

The cloud O&M center sends the fault warning (or the repair warning information) and the position of the fault device to the O&M personnel. By virtue of such information, the O&M personnel can prepare the spare parts in advance, accurately position the fault device, identify the device using a portable terminal, and repair it. After completion of repair, the O&M personnel will accurately report the type of device fault and the repair time to the cloud O&M center. The warning expert system will further improve the level of fault warning using the warning assessment model on the basis of the repair history (including evaluation of the O&M personnel), to improve the warning assessment method and increase the O&M level. Fig. 5.21 shows the procedure for repair of the fault device of the distributed resources.

The PNP O&M program based on the Internet of Things combines the technology of the Internet of Things and the O&M technology of the distributed resources, and can realize the smart O&M for lots of connected devices of the distributed resources, PNP function for O&M of the connected device, grasp

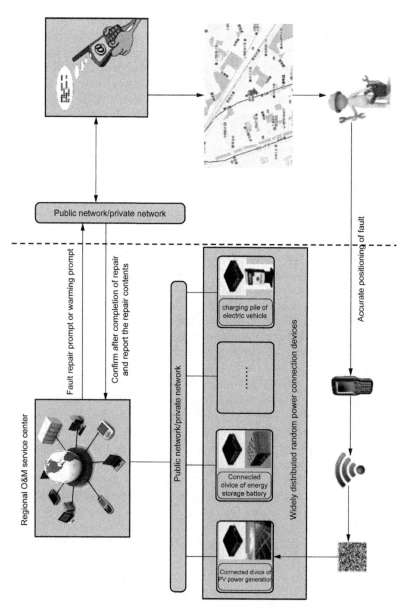

FIG. 5.20 Principle for O&M of connected device of distributed resources.

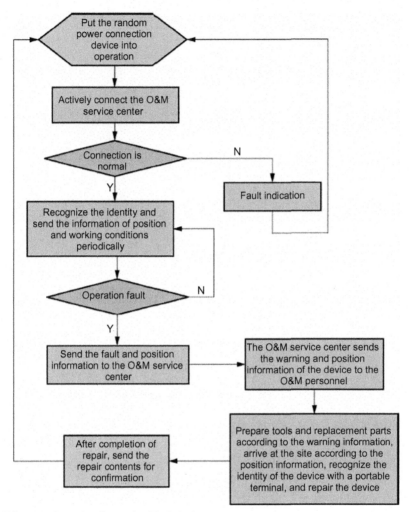

FIG. 5.21 Procedure for repair of fault device.

operation conditions of the connected devices of the distributed resources in the region in a timely and accurate manner, perform health assessment, send a warning for repair in advance, and form an electronic O&M procedure for the connected devices of the distributed devices for improving the O&M level of the distributed resources.

Chapter 6

Grid-connected operation and engineering application of distributed resources

6.1 Multifunction complementary active grid demonstration project

6.1.1 Project overview

Elion Multifunction Complementary Active Grid Demonstration Project, a national energy application technology research and engineering demonstration project, is constructed by Elion Resources Group Ltd.

The project, located in the Kubuqi Desert Park in Dugui Town, Hangjin Banner, Erdos, Inner Mongolia, at north latitude 40.67° and east longitude 106.30°, covers an area of 889 ha., which includes 114.6 ha. water area, 40.7 ha. reed wetland, 380 ha. grasslands, and 383.7 ha. desert. The area, with annual total solar radiation of 1342–1948 kWh/m^2 and annual sunshine time of 2600–3400 h, is one of the area with the highest illumination value in China.

The project is composed of a new surface power station 150 kWp, a PV shed 50 kWp, a distributed energy storage system 190 kWh, a concentrated energy storage 410 kWh, 15 pieces of hybrid charging piles, microgrid, as well as software and hardware system, civil works, and other auxiliary works of active grids. Meanwhile, the 250 kWp PV power station has been restored and grid connection and waste heat recovery of internal combustion engine generators have been reconstructed.

International Conference Center and Qixing Lake Desert Hotel are shown in Fig. 6.1. Three electric load areas are selected. Active grid systems of three types of different microgrid are constructed. The system comprises many types of distributed resources, including PV generation with installed capacity of 400 kWp, diesel generator generation with installed capacity of 1600 kW, and concentrated power storage and distributed power storage with installed capacity of 600 kWh.

Meanwhile, the demonstration system comprises many types of distributed cold and heat sources: air source heat pumps, water source heat pumps, hot-water

Distributed Power Resources. https://doi.org/10.1016/B978-0-12-817447-0.00006-7

FIG. 6.1 Qixing Lake Desert Hotel and International Conference Center. (A) (Kubuqi Desert) International Convention Center; (B) Qixing Lake Desert Hotel.

boilers, steam boilers, waste heat recovery units of internal combustion engine generators, water thermal storage and so forth. The total installed capacity is not less than 6500 kW. Besides, there are 15 AC charging piles and DC charging piles, capable of satisfying cooling, heating, and electric energy demand of more than 50,000 m² buildings and charging demand of electric vehicles.

As shown in Fig. 6.2, three microgrid node areas are arranged for above two architectural complexes of different functions: Qixing Lake Desert Hotel and International Conference Center. Here, a 911 incoming line and a 912 incoming line supply power to Microgrid #1 and #2, which mainly includes the hotel load. 912 incoming line supplies power to Microgrid #3, which mainly includes the load of the convention center.

Power generation of each distributed resource gathers at AC 0.4 kV side upon controlled conversion, supplies power to microgrids, and then will be connected into the public grid upon boosting by 0.4/10 kV step-up transformers.

Fig. 6.3 shows a functional diagram of Elion multifunction complementary active grid system. Energy management system of the active grid can conduct general monitoring, control, management, dispatching, and so forth over such links as PV, diesel generator, storage, and load. During grid-connected operation, distributed resources can be adsorbed to the greatest extent and economy of energy supply, and quality requirements for power supply can be increased. During off-grid operation, load power supply and stable operation can be assured and operation state, load of units, temperature and flow of supply and return water, temperature of incoming and outgoing water at source side, power consumption of units, and so on of heat pump units can be monitored to realize optimized energy supply.

6.1.1.1 Microgrid #1

Microgrid #1, located in Distribution Room #1 of Qixing Lake Desert Hotel, is the AC & DC hybrid microgrid, where DC bus is 750 V and AC bus is 400 V. AC bus and DC bus are connected through 100 kW DC/AC flow conditioners. AC bus is connected to low voltage side of the distribution transformer through

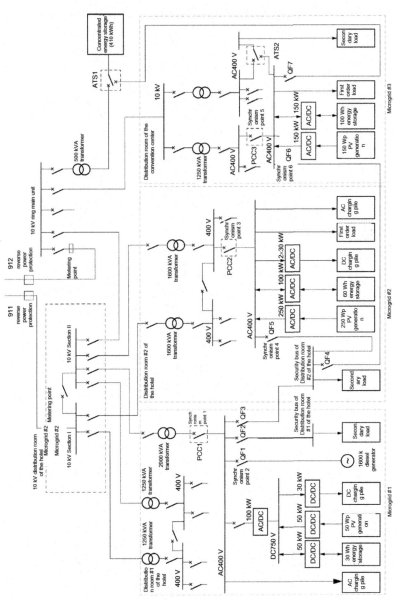

FIG. 6.2 Sketch of primary wiring of the active grid.

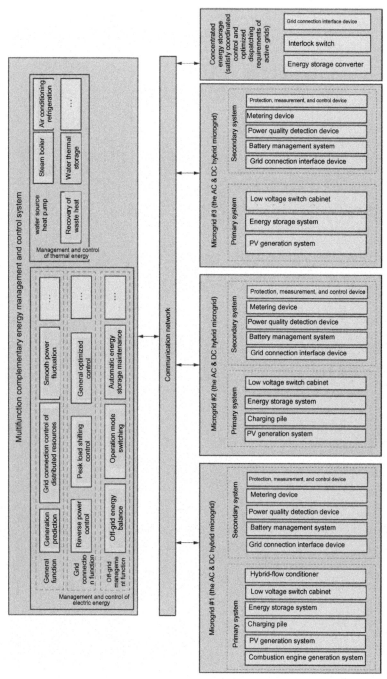

FIG. 6.3 Functional diagram of Elion multifunction complementary active grid system.

PCC1 and connected to Microgrid #2 through QF3. If the DC bus is connected with 50 kW DC/DC direct current energy storage converters, the DC bus is equipped with 30 kWh lithium batteries. If connected with 50 kW DC/DC PV converters, the DC bus is equipped with 50 kWp PV batteries. Besides, it's arranged with one 30 kW DC/DC rapid charging DC charger. Four sets of 7 kW slow charging AC charging piles are connected with AC buses and equipped with 1600 kW diesel generator units as spare power supply. Main load is composed of air conditioners, pumps, boilers, and so forth. The maximum load is 900 kW in winter and 600–700 kW in summer.

Fig. 6.4 shows a panoramic sketch of the equipment control room of Microgrid #1. Fig. 6.5 shows PV converters and energy storage converters of Microgrid #1.

FIG. 6.4 Equipment control room of Microgrid #1.

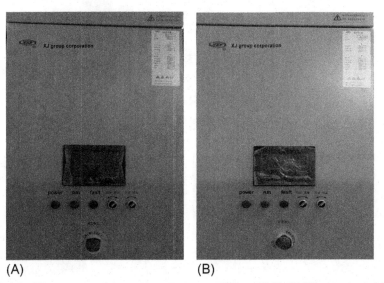

(A)   (B)

FIG. 6.5 PV converter and energy storage converter of Microgrid #1. (A) PV converter and (B) energy storage converter.

6.1.1.2 Microgrid #2

Microgrid #2, located in Distribution Room #2 of Qixing Lake Desert Hotel, is AC microgrid. The bus voltage is 400 V. It connects with the low voltage side of the distribution transformer through PCC2 and with Microgrid #3 through QF4. If the AC bus is connected with 250 kW DC/AC PV inverters, the AC bus is equipped with 250 kWp PV batteries. If the AC bus is connected with 100 kW DC/AC alternating current energy storage converters, it's equipped with 60 kWh lithium batteries, two sets of 30 kW DC/AC rapid charging DC charging generators, and eight sets of 7 kW slow charging AC charging piles. The main load is composed of hotel illumination, kitchen, firefighting, and other important loads. The load is 500 kW in summer and 200 kW in winter, including 60 kW primary load. Others are secondary loads.

Fig. 6.6 shows a panoramic sketch of the equipment control room of Microgrid #2. Fig. 6.7 shows a sketch of the energy storage converter of Microgrid #2 and the internal structure.

6.1.1.3 Microgrid #3

Microgrid #3, located in the International Convention Center, is AC microgrid. The AC bus is 400 V and it's connected with the low voltage side of the

FIG. 6.6 Equipment control room of Microgrid #2.

FIG. 6.7 Energy storage converter of Microgrid #2 and its internal structure.

FIG. 6.8   Equipment control room of Microgrid #3.

distribution transformer through PCC3. AC bus is connected with concentrated energy storage through QF7. If connected with 150 kW DC/AC PV inverters, the AC bus is equipped with 150 kWp PV batteries. If connected with 150 kW DC/AC alternating current energy storage converters, it's equipped with 100 kWh lithium batteries. The primary load of the microgrid system is 100 kW, and the secondary load is 300 kW.

Fig. 6.8 shows the equipment control room of Microgrid #3.

6.1.1.4   Concentrated energy storage

The concentrated energy storage AC bus is 400 V. If connected with 500 kW DC/AC alternating current energy storage converters, the energy storage battery will be 410 kWh lithium battery. The AC bus can be connected to 10 kV ring main units nearby the International Convention Center or Microgrid #3 via ATS1 through a step-up transformer.

6.1.2   Control of power electronic power supply

6.1.2.1   DC/AC power conversion

As for DC/AC power conversion, different control modes can be selected according to different application scenes. Refer to Table 6.1 for details.

Microgrid #1: DC/AC modules of 100 kW flow conditioner have the following main functions:

(1)   During normal connection of the AC grid, the AC microgrid and grid operate in grid-connected mode. The flow conditioner will be under DC voltage control mode (V control) to stabilize bus voltage of DC microgrid and ensure stable operation of the DC microgrid.

(2)   If the AC grid is disconnected, the flow conditioner may be under the following several working states: (1) if AC microgrid is selected to stabilize bus voltage at AC side, the flow conditioner may use V control mode to stabilize bus voltage of DC microgrid. (2) If one energy storage DC/DC module of DC microgrid is selected to stabilize bus voltage at DC side, the flow conditioner uses V/f control mode to stabilize bus voltage of AC microgrid.

TABLE 6.1 Application scene and control mode of DC/AC power modules

	Application scene	Control mode	Energy flow direction
DC/AC module	PV inverting	P/Q control	Unidirectional
	Energy storage converting	VSG control	Bidirectional
	Charging generator	P/Q control	Bidirectional
	Flow conditioner	V or V/f control	Bidirectional

Microgrid #2: DC/AC modules of 250 kW PV inverters use constant power control mode (P/Q control) for unidirectional grid connection generation. The DC/AC modules used in PV generation system form part of PV inverters. DC/AC modules of 100 kW AC energy storage converters use visual synchronous generator control mode (VSG) and form part of the energy storage converters. During grid-connected operation, power dispatching is accepted. During off-grid operation, it may operate as a voltage source. DC/AC modules of 30 kW rapid charging DC charging generators use P/Q control mode to form charging and discharging equipment so as to realize bidirectional energy flow between energy storage batteries and grid.

Microgrid #3: DC/AC modules of 150 kW PV inverters use constant power control mode (P/Q control) for unidirectional grid connection generation. DC/AC modules of 150 kW AC energy storage converters use visual synchronous generator control mode (VSG) and form part of the energy storage converters. During grid-connected operation, power dispatching is accepted. During off-grid operation, it may operate as a voltage source.

Concentrated energy storage: DC/AC modules of 500 kW AC energy storage converters use visual synchronous generator control mode (VSG) and form part of the energy storage converters. During grid-connected operation, power dispatching is accepted. During off-grid operation, it may operate as a voltage source.

6.1.2.2 DC/DC power conversion

Microgrid #1: DC/DC modules of 50 kW DC energy storage converters charge/discharge batteries under current control mode (I control) to balance system and load power. Meanwhile, it can conduct off-grid operation under voltage control mode (V control), realize rapid switching between two modes, and ensure power stability, switching rapidity and overall harmony during the switching process. DC/DC modules of 50 kW PV converters can connect with DC bus at maximum power under power control mode (P control). DC/DC modules of 30 kW rapid charging DC charging generators charge/discharge electric

TABLE 6.2 Application scene and control mode of DC/DC power modules

	Application scene	Control mode	Energy flow direction
DC/DC module	PV conversion	P control	Unidirectional
	Energy storage conversion	V/I control	Bidirectional
	Charging/discharging equipment	I control	Bidirectional

vehicles under current control mode (I control). Refer to Table 6.2 for different control modes.

6.1.2.3   Coordinated control between DC/AC modules of energy storage converters and that of PV inverters

Interconnecting technology without communication lines is used in the project to realize off-grid energy balance control of the microgrid. DC/AC modules of energy storage converters adjust drooping curve automatically according to SOC. DC/AC modules of PV inverters adjust active power automatically according to system frequency. Refer to Section 3.7.2 frequency shift control technology for the technical principle.

6.1.2.4   Control of point of common coupling

Points of common coupling of Microgrids #1, #2, and #3 are PCC1, PCC2, and PCC3, respectively. Points of common coupling are configured with the same grid connection interface devices, which are designed in low voltage switch cabinets. Refer to Figs. 6.4 and 6.6 for details. The principle of active island is introduced in Section 3.1. Grid connection interface devices can be used to protect 380 V bus. In case of failure of the microgrid, failure isolation can be realized between the microgrid and the grid. Meanwhile, its grid connection point protection within the microgrid. The following protection functions are configured: overcurrent protection, active island protection, passive island protection (over-voltage/under-voltage protection, over-frequency/under-frequency protection, etc.), and so on.

6.1.3   Control strategy for operation mode switching of multifunction complementary microgrid cluster

As for the multifunction complementary active grid demonstration project, control strategy for operation mode switching of the microgrid is formulated by the microgrid cluster energy management system and implemented through measurement and control devices, safe grid connection devices, and power electronic equipment. Safe grid connection devices, measurement and control devices, and power electronic equipment can realize switching of different operation modes of the microgrid cluster, give full play to coordination and mutual aid among different sub-microgrids, and ensure safe and reasonable work among each component of the microgrid according to the switching strategies formulated.

6.1.3.1 Control strategy for planned grid connection and off-grid mode switching

(1) Planned grid connection to off-grid mode switching

As shown in Fig. 6.2, the initial operation mode is independent grid-connected operation of three sub-microgrids. In Microgrid #1, the large grid is the main power system to supply stable AC voltage and frequency. Other distributed resources are under grid connection generation state. The grid connection switch PCC1 is switched on, and QF3 and QF4 are switched on too. Microgrid #1 supplies power to secondary load nearby through Distribution Room #2 of the hotel, and the diesel generators are closed down. In Microgrid #2, the large grid is the main power system to supply voltage and frequency support to other distributed resources that are under grid connection generation state. PCC2 is switched on, and tie switch QF5 is switched off. In Microgrid #3, the large grid is the main power system to supply voltage and frequency support to other distributed resources. PCC3 is switched on, and tie switches QF6 and QF7 are switched off. Dual power automatic transfer switch ATS2 is switched on at AC400 V bus. Microgrid #3 supplies power to secondary load through ATS2. 410 kWh concentrated energy storage can realize peak load shifting of the large grid and improve power quality through dual power automatic transfer switch ATS1 and low voltage side of 500 kVA transformer.

According to uninterruptible power supply principle for load, when implementing planned grid connection to off-grid switching, please comply with the planned grid connection to off-grid switching control logic, as shown in Fig. 6.9.

(2) Planned off-grid to grid connection mode switching

As shown in Fig. 6.2, the initial operation mode is off-grid operation of three sub-microgrids. The three sub-microgrids, operating jointly as a whole body, take diesel generators as the main power supply to support voltage and frequency of the entire microgrid and ensure stable operation of the system. Whereas for Microgrid #1, PCC1 is switched off, switches QF1, QF2, QF3, and QF4 are switched on. 100 kW flow conditioners can stabilize voltage of bus of 750 V DC side. Other distributed resources operate under current source mode through DC/DC changers. In Microgrid #2, PCC2 is switched off, On-state of QF5 synchronism can be controlled through FCK-801

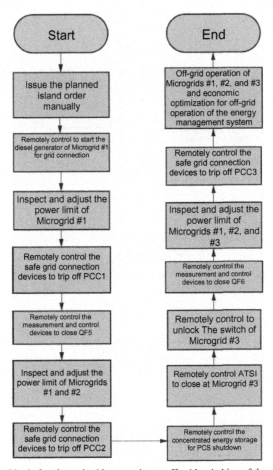

FIG. 6.9 Control logic for planned grid connection to off-grid switching of the microgrid cluster.

microcomputer-based measurement and control devices. Energy storage, PV, and AC charging generators are operating under P/Q mode at the AC side. In Microgrid #3, PCC3 is switched off; the on state of QF6 and QF7 can be controlled through FCK-801 microcomputer-based measurement and control device. The 410 kWh concentrated energy storage will supplement necessary power to the entire off-grid system through QF7 according to system requirements. ATS2 is closed at AC400 V bus.

According to orders of the main control layer or field test and the uninterruptible power supply principle for load, when implementing planned off-grid to grid connection switching, please comply with the planned off-grid to grid connection control logic shown in Fig. 6.10.

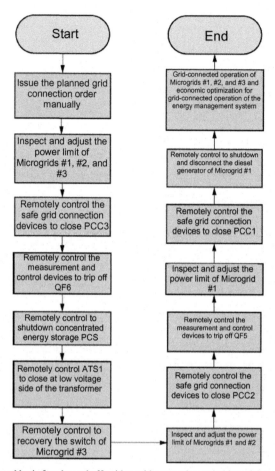

FIG. 6.10 Control logic for planned off-grid to grid connection switching of the microgrid cluster.

6.1.3.2 Control strategy for unplanned load transfer mode switching

In case of sudden failure of the partial grid, please implement unplanned load transfer mode switching according to the following principle: in case of failure of Microgrid #1, Microgrid #2 is responsible for transferring and Microgrid #3 is under grid-connected operation; in case of failure of Microgrid #2, Microgrid #1 is responsible for transferring and Microgrid #3 is under grid-connected operation; in case of failure of Microgrid #3, Microgrid #1 is responsible for transferring and Microgrid #2 is under grid-connected operation. Meanwhile, the load of Microgrids #1 and #2 can be supplied intermittently, and uninterrupted power supply will be guaranteed for load of Microgrid #3. The initial state is independent grid-connected operation of three microgrids. In case of

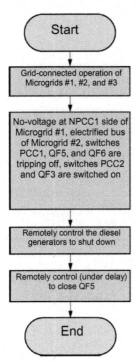

FIG. 6.11 Control logic for load transfer of Microgrid #1.

power failure at grid side of PCC1 point of Microgrid #1, implement control logic for load transfer of Microgrid #1. Refer to Fig. 6.11 for details. In case of power failure at grid side of PCC2 point of Microgrid #2, implement control logic for load transfer of Microgrid #2. Refer to Fig. 6.12 for details. In case of power failure at grid side of PCC3 point of Microgrid #3, implement control logic for load transfer of Microgrid #3. Refer to Fig. 6.13 for details.

6.1.3.3 Control strategy for unplanned load transfer recovery mode switching

Unplanned load transfer recovery mode switching can be divided into three types: unplanned load transfer recovery of Microgrid #1, unplanned load transfer recovery of Microgrid #2, and unplanned load transfer recovery of Microgrid #3. It is required that load transfer recovery is completed automatically. The load of Microgrid #1 requires interrupted power supply. Load of Microgrids #2 and #3 require uninterrupted power supply.

(1) Unplanned load transfer recovery control of Microgrid #1

Before switching, the initial operation mode is such that the PCC1 point is disconnected. Microgrid #1 and Microgrid #2 are connected to the grid through

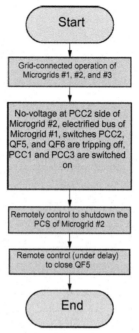

FIG. 6.12 Control logic for load transfer of Microgrid #2.

PCC2. Microgrid #2 supplies power to the load of Microgrid #1 though tie switches QF3 and QF5. Microgrid #3 is connected to the grid independently. Diesel generators are not started. The 50 kW DC energy storage converters in Microgrid #1 are operating under current source mode. The 100 kW AC energy storage converters in Microgrid #2 and 150 kW AC energy storage converters in Microgrid #3 use visual synchronous generator technology, making sure that energy storage converters are provided with inertia and damping of simulated traditional synchronous generators and participating in primary frequency modulation and primary voltage regulation of the grid. If the side of PCC1 point of Microgrid #1 is electrified, please implement the control strategy for unplanned load transfer recovery. Refer to Fig. 6.14 for the control logic.

(2) Unplanned load transfer recovery control of Microgrid #2

Before switching, the initial operation mode is such that the PCC2 point is disconnected. Microgrid #1 and Microgrid #2 are connected to the grid through PCC1. Microgrid #1 supplies power to load of Microgrid #2 though tie switches QF3 and QF5. Microgrid #3 is connected to the grid independently. Diesel generators are not started. If the side of PCC2 point of Microgrid #2 is electrified, please implement the control strategy for unplanned load transfer recovery. Refer to Fig. 6.15 for the control logic.

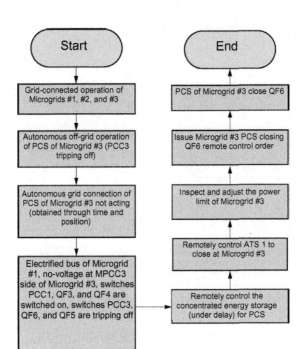

FIG. 6.13 Control logic for load transfer of Microgrid #3.

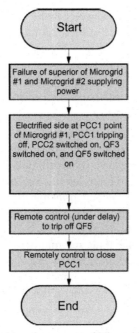

FIG. 6.14 Control logic for load transfer recovery of Microgrid #1.

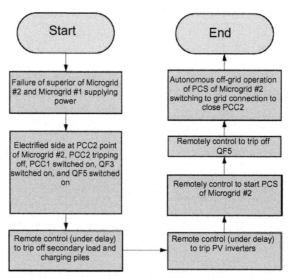

FIG. 6.15 Control logic for load transfer recovery of Microgrid #2.

(3) Unplanned load transfer recovery control of Microgrid #3

Before switching, the initial operation mode is such that the PCC3 point is disconnected. Microgrid #1 and Microgrid #3 are connected to the grid through PCC1. Microgrid #1 supplies power to load of Microgrid #3 though tie switches QF4 and QF6. Microgrid #2 is connected to the grid independently. Diesel generators are not started. If the side of PCC3 point of Microgrid #3 is electrified, please implement the control strategy for unplanned load transfer recovery. Refer to Fig. 6.16 for the control logic.

6.1.3.4 Control strategy for unplanned grid connection to off-grid mode switching

Unplanned grid connection to off-grid control mode refers to situations in which the entire microgrid cluster is forced to implement grid connection to off-grid mode according to certain principles, due to simultaneous power loss at the side of the grid connection point of three sub-microgrids as a result of failure of the large grid. The design mode switching principle is as follows: interrupted power supply is allowed for Microgrid #1 and Microgrid #2, and uninterrupted power supply will be guaranteed for Microgrid #3.

The initial state is such that three sub-microgrids are subject to independent grid-connected operation. QF5 and QF6 are disconnected. In case of power loss of the large grid, Microgrid #3 will complete seamless switching between grid connection and off-grid, first. Now, Microgrids #1 and #2 are under power failure. Second, Diesel Generator #1 is started as the main power supply. Microgrid #2 controls QF5 through FCK801-microcomputer-based measurement and

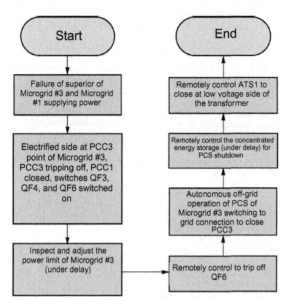

FIG. 6.16 Microgrid #3 load transfer recovery control logic.

control devices to realize grid connection power supply. The backstage of Microgrid #3 will send synchronous closing order to 150 kW energy storage converters, which will use presynchronization to control QF6 synchronization grid connection after receiving the order. Finally, the diesel generator will be used as a main power supply, and off-grid operation of three microgrids will be realized. Refer to Fig. 6.17 for the control logic.

6.1.3.5 Control strategy for unplanned grid connection to off-grid recovery mode switching

Unplanned grid connection to off-grid recovery control refers to automatic switching of the entire microgrid from off-grid mode to grid connection mode when the large grid is electrified. The design mode switching principle is as follows: interrupted power supply is allowed for Microgrid #1, and uninterrupted power supply will be guaranteed for Microgrids #2 and #3.

The initial state is that three sub-microgrids operate under off-grid. PCC1, PCC2, and PCC3 are disconnected, and tie switches QF5 and QF6 are switched on.

When the large grid is electrified, first, iSI-810 safe grid connection device controls the PCC2 to be on and realizes off-grid to grid-connected operation of the entire microgrid. Second, disconnect the combined switch to realize independent grid-connected operation of Microgrid #2. Third, the backstage controls microcomputer-based measurement and control

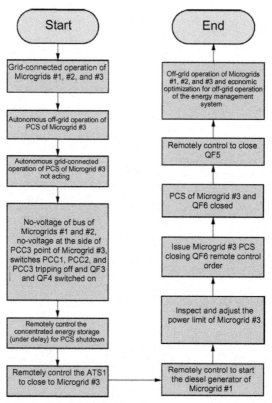

FIG. 6.17   Control logic for unplanned grid connection to off-grid switching of the microgrid cluster.

devices to trip off QF6. Microgrid #3 is subject to independent off-grid operation. 150 kW energy storage converter is used as the main power supply. Fourth, iSI-810 safe grid connection device controls PCC3 to be on to the realize off-grid to grid-connected operation of the Microgrid #3 system. Fifth, iSI-810 safe grid connection device controls PCC1 to be on to realize off-grid to grid-connected operation of the Microgrid #1 system. Refer to Fig. 6.18 for detailed control logic.

6.1.4   Engineering application and its guiding meaning

The jointed supply and complementary coordination of energy can improve energy transmission and utilization efficiency; the integration of energy requirements, development diversity, and electric system, is the basic platform for power development, configuration, and utilization, effectively; and can give full play to core configuration roles of the grid in energy supply; coordinate

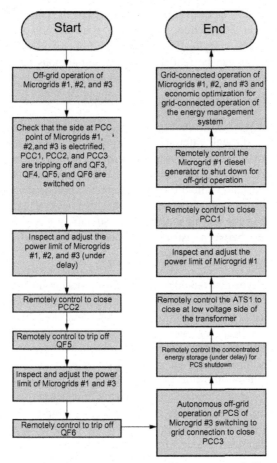

FIG. 6.18 Control logic for unplanned off-grid to grid connection switching of the microgrid cluster.

various forms of energy supply through the grid to realize complementary operation; and bring comprehensive benefits into play. The idea of multifunction complementation is used to modify the grid, making it an energy supply and demand network integrating energy acquisition, energy transmission, energy storage and distribution, and cold and heat conversion and transmission.

6.1.4.1 Application of key technologies

(1) Modulated paralleling technology

The distributed PV, wind turbines, energy storage, charging generators, and so on are connected into the grid through power electronic converters. Standard three-level DC/DC and DC/AC power modules are used in the converters to form conversion devices of different capacity from the perspective of convenient maintenance. Meanwhile, appropriate control software is embedded according to different application scenes. As for power modules, different access requirements can be realized through autonomous paralleling, which can reduce the types of devices and facilitate maintenance.

(2) Injection-type active island detection technology

The 20 Hz component equivalent to zero sequence component is injected into the 380 V system through external low frequency power modules with small power. Islands are identified according to change characteristics of 20 Hz component before and after occurrence of an island so as to achieve an operation state of an active island without relying on inverters and communication, solve detection dead zone and slow speed (easily causing issues related to power quality and other issues in existing methods), and satisfy electrical interlocking safety requirement between distributed resources and the grid.

(3) Virtual synchronous generator technology

Power electronic type distributed resources use digital circuit for control, which is featuring with rapid transient response speed. Besides, the distributed resources are unable to participate in frequency regulation and voltage regulation of the grid. Self-approximate optimization virtual synchronous generator technology makes power electronic type distributed resources capable of adjusting inertia through self-adaptation according to turbulence in system power, so as to prevent too slow or too rapid system dynamic response caused by excessive large or small rotary inertia and too long or too short transient state process caused by excessive large or small damping, improve grid connection stable region of the new energy power generation system, and realize "grid-friendly" characteristics and networking characteristics of converters.

(4) P/U control technology

P/U control technology can be used to make distributed resources adjust output automatically according to voltage to solve the problem of distributed generation caused by voltage rise due to excessive active power and improve the generating capacity penetration of distributed resources.

(5) Microgrid autonomous operation control technology without communication lines

Instead of relying on communication, information transfer is conducted through fundamental frequency regulation of the microgrid. Besides, control equipment is not added, and concentrated control units of the microgrid are canceled so as to realize autonomous stable paralleling operation between the energy storage system and distributed generation, establish a microgrid with simple physical structure, high reliability, and low cost, and satisfy plug-and-play commercial application of the microgrid.

(6) Presynchronization grid connection technology

Off-grid to grid connection of the microgrid cannot ensure completely consistent voltage amplitude and phase at both sides. When closing, a huge impact current may make grid connections fail and damage equipment. Presynchronization grid connection technology can ensure consistent voltage amplitude and phase of the microgrid and that of the grid through gradually approaching amplitude and phase, realizing "zero impact" grid connection.

6.1.4.2 Multifunction complementary energy management

The multifunction complementary active grid energy management system can realize orderly, complementary, cascading, and optimized utilization of conventional energy resources and renewable energy resources at the supply side, and achieve satisfy electric energy and thermal energy dispatching requirements, and achieve multifunction complementation from generation end and consumption end of distributed resources through collecting and controlling information flow of many types of energy sources. The interface of Elion multifunction complementary active grid energy management system is shown in Fig. 6.19. Mutual coordination between power supply and heat supply can be used to improve the utilization ratio of renewable energy generation and to realize economic operation of energy supply.

Power supply: Efforts will be made to improve utilization ratio of distributed resources utilizing electric dispatching, energy storage, and other means guaranteed by the large grid power supply under the premise of satisfying power safety and quality requirements of customers. When the generation power of distributed resources and electric load of the microgrid are higher than the set ratio, the energy dispatching system will first store surplus quantity of electricity into energy storage and batteries of electric vehicles through electric energy dispatching. After complete charging of batteries, operation load of heat pumps will be improved to store surplus electric energy into hot (cold) load of buildings through heat (cooling) quantity. When the hot (cold) load of buildings reach set value, if generation power of distributed resources is still larger than electric load of the grid, power output of distributed resources will be limited. During the overall process, if the power of distributed resources is lower than electric load, stored energy dispatching will be stopped.

Heat supply: Heat supply is mainly composed of supply of such four energies as waste heat, solar energy, electric energy, and fuel. The dispatching of various heat sources is optimized to utilize waste heat and solar energy, first. Then, water source heat pumps are used, followed by fuel boilers. Waste heat is a kind of free energy that can only be generated during the start-up of diesel generators. Solar energy is renewable energy, which can only be generated in daytime with clear weather. Both types of energy are intermittent unstable energy and cannot be used as a main heat source for cold and heat load separately. Water source heat pumps can provide much thermal energy to buildings

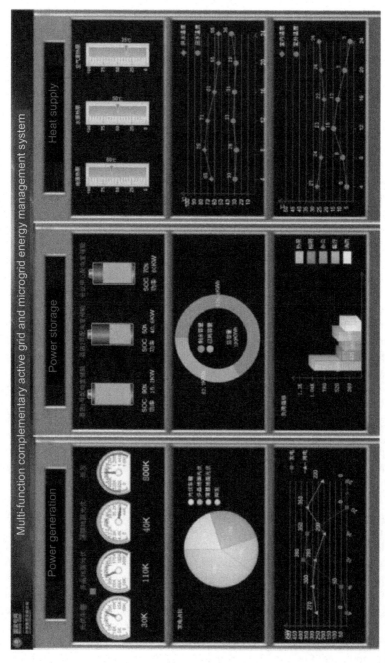

FIG. 6.19 Elion multifunction complementary active grid energy management.

through consuming little electric energy, which can be used as the main source for the heating system to ensure basic heat supply requirements of buildings. Boilers can generate thermal energy through combusting diesel oil. Boilers can be used as peak shaving and emergency heat sources and be started in case of failure or power shortage of other heat sources.

6.1.4.3   Engineering application and its guiding meaning

Multifunction complementary active grid can give full play to the configuration role of the grid in energy supply to realize many types of energy input (solar energy, natural gas, willow geothermal energy, etc.), many types of energy supply output (cold, heat, electricity, etc.), and many energy transfer units (PV, gas turbines, internal combustion engines, energy storage systems, etc.); turn the grid from the one with electric energy distribution only into a new type multiple energy network system integrating electric energy collection, electric energy transmission, electric energy storage, and distribution, as well as transfer and transmission of cold and heat; make the best of renewable energy; and ensure electric energy quality, operation stability, and power supply reliability of the system, as well as comprehensive supply of regional cold and heat, which is of demonstration and promotion value.

6.2   High-density distributed energy resources connecting into the AC & DC hybrid microgrid demonstration project

6.2.1   Project overview

The demonstration project, based on national 863 topic "High-density distributed energy resources connecting into the AC & DC hybrid microgrid key technology," is organized by Zhejiang University, Tianjin University, China Electric Power Research Institute, Hefei University of Technology, North China Electric Power University, Xuji Group Corporation, Beijing Siwind Turbineg Automation Co., Ltd., and other scientific research institutes and equipment manufacturers, focusing on solving such key technologies as optimized configuration, stable control, comprehensive protection, power quality treatment, and energy optimization of network structure of high-density distributed energy connecting into the AC & DC hybrid microgrid system.

   The project, focused mainly on existing PV power stations within the plant area of Zhejiang Century Huatong Group Co., Ltd., is equipped with a wind power generation system and energy storage system with appropriate capacity. And the purpose is to build a grid connection type low voltage AC & DC hybrid microgrid structure. During grid-connected operation, multiple coordinations and interactions between microgrid and grid can be realized. During off-grid operation, maximum utilization of distributed energy resources can be realized based on ensuring the power supply of important loads. The existing solar energy generation capacity in the plant area is about 2.4 MWp. The roof solar energy subsystem is connected into 10 kV bus of the generation distribution station of the PV system through cable shafts and cable trenches upon local inversion. The power supply in the plant area is connected through a 10 kV Huatong 940 dedicated line of Shangyu 110 kV Guoqing Substation. The 10 kV line of the Century Huatong plant area is reduced to 0.4 kV through seven sets of 1250 kVA transformers to supply power to the Huatong plant area. The plant area low voltage side 0.4 kV is a direct grounding system.

The AC & DC hybrid microgrid composed of high-density distributed energy resources constructed in the demonstration project is a grid connection type microgrid project composed of many types of distributed resources, including total capacity of distributed PV reconstructed or accessed 2.4 MWp, two sets of newly constructed 5 kW wind power generation systems, one set of 250 kW/ lMWh lead-acid battery energy storage system, one set of 20 kW/50 kWh liquid flow battery energy storage system, and one set of the AC & DC hybrid microgrid power conversion and network system. The maximum AC load in the system is 1.2 MW. The DC load comprises 10 sets of 50 kW injection molding machines, 50 kW LED head lamps, and 8 sets of DC charging piles. And the maximum DC load is about 1 MW.

Refer to Fig. 6.20 for an effect drawing of the project layout. Fig. 6.21 shows a project introduction and a picture of the energy storage equipment site.

Fig. 6.22 shows a primary wiring diagram of Shangyu Demonstration Station in the AC & DC hybrid microgrid. The system includes 10 kV AC bus and 560 V DC bus. Both the 10 kV AC bus and 560 V DC bus use sectionalized configuration, dividing the power distribution area into Section I of 10 kV AC bus, Section II of 10 kV AC bus, Section I of 560 V DC bus, and Section n of 560 V DC bus, where 10 kV AC bus is connected to the Shangyu 10 kV public network. Section I of 10 kV AC bus and Section n of 10 kV AC bus are connected through tie switch KG1. Section I of 560 V DC bus and Section n of 560 V DC bus are connected through DC circuit breakers.

(1) Section I of 10 kV AC bus

Section I of AC bus is connected with 520 kWp PV power generation system, and two sets of 250 kVA flow conditioners and AC load. 520 kWp PV battery is connected into Section I of 10 kV AC bus after inverting by 500 kW PV inverters and boosting by one set of 630 kVA step-up transformer. The 10 kV AC load is about 600 kW. Section I of 10 kV AC bus and Section I of 560 V DC bus are connected through two sets of 250 kVA flow conditioners.

(2) Section II of 10 kV AC bus

Section II of AC bus is connected with one set of 250 kVA flow conditioner, one set of 250 kVA DC transformer, and one set of AC power quality treatment device and AC load. 520 kWp PV battery is connected into Section II of 10 kV AC bus after inverting by 500 kW PV inverters and boosting by one set of 630 kVA step-up transformer. The AC load is about 600 kW.

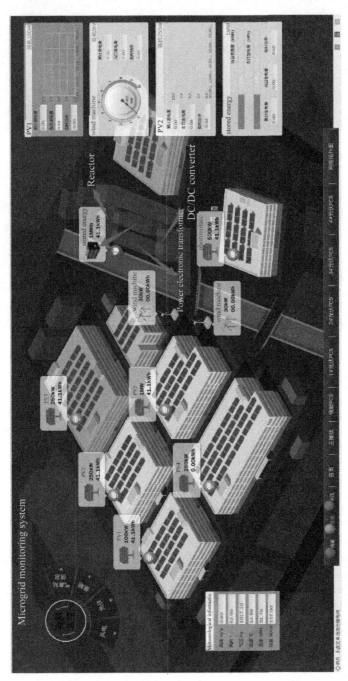

FIG. 6.20 Effect drawing of project layout.

(A)

(B)

FIG. 6.21 Project introduction and picture of energy storage equipment site. (A) Front and (B) back.

Section II of 10 kV AC bus and Section II of 560 V DC bus are connected with one set of DC transformer through one set of 250 kVA flow conditioner. An AC power quality treatment device is connected into Section n of 10kv AC bus after boosting by one set of 630kVA power distribution transformers.

(3) Section I of 560 V DC bus

Section I of DC bus is connected with 20 kW DC bidirectional DC/DC energy storage converters and energy storage batteries are 50 kWh liquid flow batteries. PV #1 is connected with 100 kW PV DC/DC converters, and solar panels are of 96kWp. PV #2 is connected with 250 kW PV DC/DC converters, 235 kWp solar panels, and one set of 50A DC power quality treatment devices. DC load includes eight sets of 60 kW DC charging piles and five sets of 50 kW injection molding machines in Plant #4.

(4) Section II of 560 V DC bus

Section II of DC bus is connected with two sets of 5 kW AC/DC wind power converters, two sets of 250 kW PV DC/DC converters, 540 kWp solar panels, one set of 250 kW DC bidirectional energy storage converters with lMWh lead

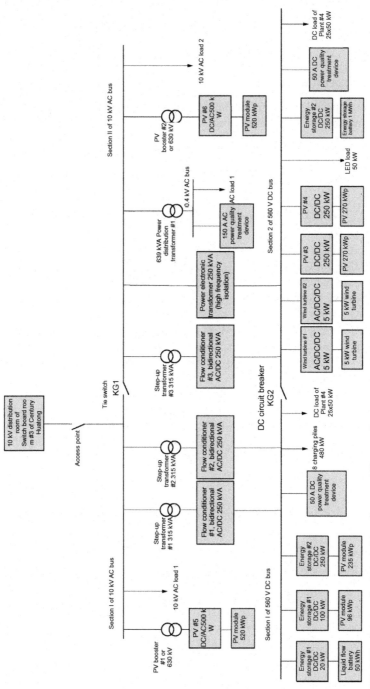

FIG. 6.22 Primary simulation wiring diagram of Shangyu Demonstration Station in the AC & DC hybrid microgrid.

acid battery, and one set of 50A DC power quality treatment devices. And DC load is 50 kW LED lighting load and five sets of 50 kW injection molding machine in Plant #4.

6.2.2 Operation mode

To fully display flexibility of operation mode of the AC & DC hybrid microgrid, to ensure continuous and highly reliable power supply of normal part in case of failures, and to realize highly effective and reliable access and adsorption of high-density distributed resources, the AC & DC hybrid microgrid in the demonstration project is designed with four operation modes. Refer to Table 6.3 for details.

6.2.2.1 Grid-connected operation mode (both KG1 and KG2 are closed)

Flow conditioners #1 and #2 use DC side droop control to stabilize the voltage of the DC bus. Flow conditioner #3 uses AC side droop control (Now, the voltage and frequency of Section II of 10 kV AC bus is actually stabilized by the large grid. However, to make sure that no mode switching is required for Flow conditioner #3 when KG1 is disconnected, AC side droop control is adopted.) and DC transformers use AC P/Q control. Meanwhile, an appropriate algorithm is adopted to realize current sharing control between three sets of flow conditioners and DC transformers. The energy storage system, which is used as a controlled source, can realize SOC control.

Firstly, DC microgrid in the lowest layer will be taken into consideration. Maximum generating capacity of distributed resources in DC microgrid #1 on the left side of the DC circuit breaker is 350 kW, the energy storage is 20 kW, and total load is 730 kW. Maximum generating capacity of distributed

TABLE 6.3 Relationship between KG1 and KG2 states and operation mode

Operation mode	KG1	KG2	Operation mode	KG1	KG2
Grid-connected operation	Closed	Closed	Subsection operation of AC bus	Disconnected	Closed
Subsection operation of DC bus	Closed	Disconnected	AC & DC hybrid microgrid isolated operation	Disconnected	Disconnected

resources in DC microgrid #2 on the right side of the DC circuit breaker is 510 kW, the energy storage is 250 kW, and total load is 300 kW.

As it is under grid connection state, to ensure energy storage is under a relatively high SOC state, the energy storage state will absorb the power or power balance state. Now, maximum generating capacity of distributed resources of the entire DC microgrid is 860 kW, the energy storage is 270 kW, and maximum total load is 800 kW.

As for 10 kV AC bus, a 10 kV AC microgrid and a 500 kW PV system with maximum generating capacity of 500 kW are provided on both sides of KG1. Besides, there are two 10 kV AC loads and one 0.4 kV AC load.

6.2.2.2 Subsection operation mode of DC bus (KG1 closed, KG2 disconnected)

Now, Flow conditioners #1 and #2 use DC side droop control to stabilize voltage of DC bus #1. The voltage of DC bus #2 is stabilized through its internal Energy storage #2. Flow conditioner #3 uses AC side droop control, so as to realize current sharing control with the DC transformer, which is used as a P/Q controlled source.

Firstly, DC microgrid in the lowest layer will be taken into consideration. Maximum generating capacity of distributed resources in DC microgrid #1 on the left side of the DC circuit breaker is 350 kW, the energy storage is 20 kW, and total load is 730 kW. Now, as DC microgrid #1 is still under grid connection state, Flow conditioners #1 and #2 will transmit power through current sharing control. DC microgrid #2 on the right side of the DC circuit breaker is free from flow conditioners or DC transformers. (Flow conditioner #3 uses AC side droop control, unable to stabilize DC voltage. The DC transformer is under P/Q control, unable to stabilize DC voltage as well.) Therefore, only energy storage batteries inside can be used for stabilization control. The maximum generating capacity of distributed resources of Microgrid #1 is 350 kW, the energy storage is 20 kW, and maximum total load is 730 kW. Maximum generating capacity of distributed resources of DC microgrid #2 is 510 kW, the energy storage is 250 kW, and maximum total load is 300 kW.

On the 10 kV AC bus, voltage and frequency of the bus of the entire 10 kV AC microgrid is stabilized by the large grid due to closure of KG1. 10 kV AC microgrid and PV system with maximum generating capacity of 500 kW are arranged on both sides. Besides, there are two 10 kV AC load and one 0.4 kV AC load.

6.2.2.3 Subsection operation mode of AC bus (KG1 disconnected, KG2 closed)

The voltage of the entire DC bus is controlled by Flow conditioners #1 and #2, which are subject to paralleling operation to realize current sharing between two conditioners. Flow conditioner #3 uses AC side droop control to stabilize voltage and frequency of AC bus #2, which can realize current sharing control with the DC transformer, and is used as a P/Q controlled source. The energy storage system, which is used as a controlled source, can realize SOC control.

First, the DC microgrid in the lowest layer will be taken into consideration. Maximum generating capacity of distributed resources in DC microgrid #1 on the left side of the DC circuit breaker is 350 kW, the energy storage is 20 kW, and total load is 730 kW. Maximum generating capacity of distributed resources in DC microgrid #2 on the right side of the DC circuit breaker is 510 kW, the energy storage is 250 kW, and total load is 300 kW. As it is under grid connection state, to ensure that energy storage is under a relatively high SOC state, the energy storage state will absorb power or attain a power balance state. Then the maximum generating capacity of distributed resources of the entire DC microgrid is 860 kw, the energy storage is 270 kW, and maximum total load is 1030 kW.

On the 10 kV AC bus, voltage and frequency of the bus of the 10 kV AC microgrid #1 on the left side is stabilized by the large grid due to the disconnection of KG1. It includes a PV system with maximum generating capacity of 500 kW and a 10 kV AC load. The voltage and frequency of the bus of the 10 kV AC microgrid #2 on the right side is stabilized by Flow conditioner #3, which includes one PV with maximum generating capacity of 500 kW, one 10 kV AC load, and one 0.4 kV AC load.

6.2.2.4 AC&DC hybrid microgrid isolated operation mode (both KG1 and KG2 are disconnected)

Due to disconnection of **KG**1 and **KG**2, the entire the AC & DC hybrid microgrid is divided into two parts: 10 kV AC microgrid #1 and DC microgrid #1 under paralleling operation and 10 kV AC microgrid #2 and DC microgrid #2 under isolated operation. The voltage of DC bus #1 is controlled by Flow conditioners #1 and #2, which are subject to paralleling operation to realize current sharing between two conditioners. Voltage of DC bus #2 is stabilized and controlled by Energy storage #2. The voltage and frequency of the bus of 10 kV AC microgrid #2 are stabilized through droop control at AC side of Flow conditioner #3. DC transformer #4, as PQ controlled source or AC droop control, can realize current sharing control with Flow conditioner #3.

First, DC microgrid in the lowest layer will be taken into consideration. Maximum generating capacity of distributed resources in DC microgrid #1 on the left side of the DC circuit breaker is 350 kW, the energy storage is 20 kW, and total load is 730 kW. Maximum generating capacity of distributed resources in DC microgrid #2 on the right side of the DC circuit breaker is 510 kW, the energy storage is 250 kW, and total load is 300 kW. As DC microgrid #1 is under grid connection state, to ensure energy storage is under a relatively high SOC state, the Energy storage #1 state will absorb power or power balance state. DC microgrid #2 is under isolated operation state. Energy storage

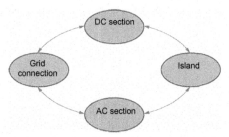

FIG. 6.23   Diagram for switching operation modes of the AC & DC hybrid microgrid.

#2, as the main power supply for the entire island system, can stabilize the voltage of DC microgrid #2.

On the 10kV AC bus, voltage and frequency of the bus of the 10kV AC microgrid #1 on the left side is stabilized by the large grid due to disconnection of KG1. It includes a PV system with maximum generating capacity of 500kW and a 10kV AC load. The voltage and frequency of the bus of the 10kV AC microgrid #2 on the right side is stabilized by Flow conditioner #3, which includes one PV with maximum generating capacity of 500kW, one 10kV AC load, and one 0.4kV AC load.

Fig. 6.23 shows a diagram for switching operation modes of the AC & DC hybrid microgrid.

6.2.3   Control of power electronic power supply

Flow conditioners #1, #2, and #3 and DC transformers in the AC & DC hybrid microgrid are connected with AC buses and DC buses to realize energy exchange and mutual support between the AC microgrid and DC microgrid.

6.2.3.1   Flow conditioner

Flow conditioners may select different control modes according to different operation modes. Refer to Fig. 6.24 for flow conditioners on the project site.

Flow conditioners #1 and #2 are operating under grid connection state, using DC droop control mode to stabilize bus voltage of DC microgrid and realize stable operation and power distribution of the DC microgrid.

Adjusting and controlling reference voltage can change output voltage to recover DC bus voltage. Adjusting the droop coefficient can change the power distribution ratio between Flow conditioners #1 and #2 to ensure precise sharing between power of Flow conditioners #1 and #2.

Flow conditioner #3, using AC droop control, features the following functions:

(1)   When the system operates under grid-connected operation state and DC bus is subject to subsection operation, Flow conditioner #3 can adjust reference active power and reference reactive power to change output power

FIG. 6.24 AC & DC hybrid microgrid flow conditioners on the project site.

of flow conditioners and to realize power sharing between the AC microgrid and DC microgrid.
(2) When the system operates under AC bus subsection mode and island operation mode, Flow conditioner #3 will adjust and refer to voltage amplitude and reference power to stabilize voltage and frequency of AC bus of AC microgrid and to ensure stable operation of the AC microgrid. Refer to Table 6.4 for details.

6.2.3.2 DC transformer

Refer to Fig. 6.25 for a picture of the project site of the DC transformer.
The DC transformer is mainly provided with the following functions:

(1) When system operates under grid-connected operation mode and DC bus subsection operation mode, DC transformers use P/Q control mode to realize power dispatching at the AC side and DC side.
(2) When system operates under AC bus subsection mode and island operation mode, DC transformers use AC droop control mode, to realize AC voltage droop control with Flow conditioner #3 and shared power control. Refer to Table 6.5 for details.

6.2.3.3 Energy storage converter

The energy storage converter mainly has the following functions:

(1) When the system operates under grid-connected operation mode and AC bus subsection operation mode, energy storage DC/DC converters can charge/discharge batteries through power control mode (P control).

TABLE 6.4 Control mode of flow conditioners

Equipment description	Control mode of converters	Adjustable parameter	Function
Flow conditioners #1 and #2	DC V/I droop	Reference voltage	Stabilize voltage of DC bus
		Droop coefficient	Change power sharing ratio
		Reference current	Adjust current and realize power sharing without difference
Flow conditioner #3	P/f and V/Q droop	Reference amplitude	Stabilize voltage amplitude of AC bus
		Reference frequency	Stabilize frequency of AC bus
		Active droop coefficient	Adjust active droop coefficient
		Reference active power	Adjust active power of Flow conditioner #3
		Reactive droop coefficient	Adjust reactive droop coefficient
		Reference reactive power	Adjust reactive power of Flow conditioner #3

FIG. 6.25 Project site of DC transformers.

TABLE 6.5 Control mode of DC transformers

Equipment description	Control mode of converters	Adjustable parameter	Function
DC transformer	P/Q control	Reference active power	Active power controllable
		Reference reactive power	Reactive power controllable
	P/f and V/Q droop	Reference amplitude	Stabilize voltage amplitude of AC bus
		Reference frequency	Stabilize frequency of AC bus
		Active droop coefficient	Adjust active droop coefficient
		Reference active power	Adjust active power of DC transformer
		Reactive droop coefficient	Adjust reactive droop coefficient
		Reference reactive power	Adjust reactive power of DC transformer

Meanwhile, as controlled source, accept power transfer to realize energy storage SOC control.

(2) When the system operates under DC bus subsection mode and island operation mode, energy storage DC/DC converters use voltage control mode (V control) to stabilize voltage of DC bus. Refer to Table 6.6 for details.

6.2.3.4   PV DC/DC converter

PV DC/DC converters use maximum power (MPPT, maximum power tracking) output to realize maximum PV output. Meanwhile, it can accept power dispatching of the energy management system. Refer to Table 6.7 for details.

6.2.4   Energy management system

6.2.4.1   MEMS-8500 AC & DC hybrid microgrid energy management system

As for the AC & DC hybrid microgrid AC & DC hybrid flow sections, the AC & DC hybrid microgrid can operate under four operation modes according to

TABLE 6.6 Control mode of energy storage DC/DC converters

Equipment description	Control mode of converters	Adjustable parameter	Function
		Reference voltage	Stabilize voltage of DC bus
Energy storage DC/DC converter	DC droop control	Reference current	Adjust output current and realize power sharing without difference. Rated power charging and discharging will make SOC at normal level
		Droop coefficient	Change power shareing ratio

TABLE 6.7 Control mode of PV DC/DC converters

Equipment description	Means of control of converters	Adjustable parameter	Function
PV DC/DC converter	MPPT/constant power (P control)	Power dispatching order	If adjustable constant power P is larger than maximum PV prediction power, maximum prediction power will prevail. If adjustable constant power P is smaller than maximum PV prediction power, constant power P output will prevail

states of AC tie switches and DC circuit breakers: grid-connected operation, AC subsection operation, DC subsection operation, and off-grid operation. Fig. 6.26 shows the main wiring diagram of the energy management system of the AC & DC hybrid microgrid.

The entire AC & DC hybrid microgrid can be deemed as a primary microgrid. The grid on the right side of AC tie switches and switches of DC circuit breakers can be deemed as AC Section II sub-microgrid, which is a subordinate microgrid of the AC & DC hybrid microgrid. The DC microgrid on the right side of the DC circuit breaker will be deemed as DC Section II sub-microgrid. The DC Section II sub-microgrid is a subordinate microgrid of AC Section II sub-microgrid.

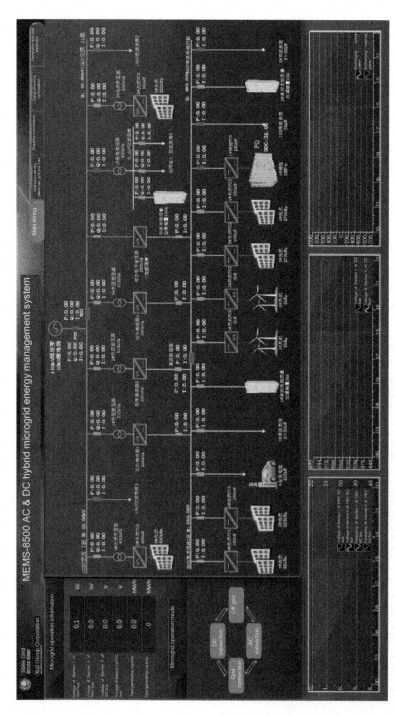

FIG. 6.26 Main wiring diagram of the AC & DC hybrid microgrid energy management system.

During grid-connected operation and AC subsection operation, the AC & DC hybrid microgrid is under grid-connected state. The AC & DC hybrid microgrid energy management system is featured with such functions as interchange power control and smooth output control.

During DC subsection operation, DC Section II sub-microgrid is under off-grid operation mode. The AC & DC hybrid microgrid energy management system is featuring with off-grid energy balance of DC Section II sub-microgrid. The grid connection part of the grid is still featuring with power exchange control and smooth DG output.

During off-grid operation, the AC & DC hybrid microgrid energy management system is featuring with off-grid energy balance function of AC Section II sub-microgrid. The grid connection part of the grid is still featured with power exchange control and smooth DG output functions.

6.2.4.2 Description of business functions

(1) Interactive optimization of resources, grid, and load of the microgrid

Distributed generation in the AC & DC hybrid microgrid is complemented with large grid power supply. Exchanging power with the large grid is a general operation mode of the AC & DC hybrid microgrid. In the AC & DC hybrid microgrid, the resources, grid, and load refer to distributed resources, grid, and load respectively. The development and investment of sources, grid, and load are managed and controlled by different bodies and belong to different bodies. In the power market, distributed resources, generation bodies, grid enterprises, and power consumers are independent entities with respective benefits. They have independent decision-making authority and contact with each other through quantity of electricity or electrovalency. Due to different investment objects and focuses, investment benefits of different parties are affected by the decisions of the other parties. The aforementioned three parties have formed a game relation. It is hopeful that multiobject optimization game theory against complicated bodies can solve the problem. In case of load fluctuation in the microgrid, the AC & DC hybrid microgrid energy management system will compare generation cost of the microgrid and purchase (sales) cost of the large grid, use chaotic particle swarm optimization based on the multiple party game model, and optimize and determine the adjustment amount of each distributed resource output and purchasing (selling) quantity to the large grid, which can ensure power balance in the AC & DC hybrid microgrid; provide reliable, high quality, and economic electric energy to users; realize optimized energy structure; and ensure coordinated development among distributed resources, grid, and load. Fig. 6.27 shows interactive optimization interface of resources, grid, and load of the AC & DC hybrid microgrid energy management system.

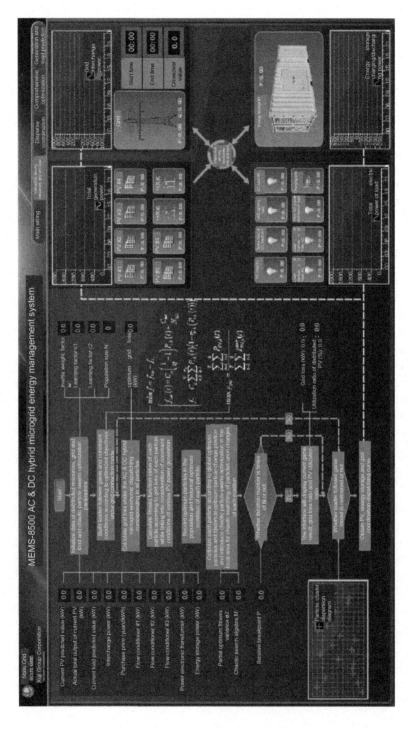

FIG. 6.27 Interactive optimization among resources, grid and load of the AC & DC hybrid microgrid energy management system.

(2) Disperse and interactive coordination between AC & DC hybrid flow sections

The AC & DC hybrid microgrid is comprised of an AC sub-microgrid and DC sub-microgrid. The AC area and DC area are connected through AC/DC bidirectional flow conditioners. Multiple sets of paralleling AC/DC bidirectional flow conditioners form a AC & DC hybrid flow section, playing an important role in cross-regional power interaction and maintaining dynamic power balance in the AC & DC hybrid microgrid.

In the AC & DC hybrid microgrid, mutual support and interconnection between AC area and DC area can be realized through bidirectional power flow. Respective power balance in AC area and DC area are completed jointly by load, distributed generation units, and AC/DC bidirectional flow conditioners. Therefore, AC/DC bidirectional flow conditioners are responsible for power exchange between AC area and DC area to reflect the interaction of active power. Considering fluctuation and uncertainty of output of distributed resources and strong randomness of AC & DC load, research on power flexibility of AC & DC hybrid flow sections will be conducted. Therefore, the AC & DC hybrid microgrid energy management system will develop such advanced function application as disperse and interactive coordination between AC & DC hybrid flow sections, which can realize cross-area flow complementation between hybrid sub-microgrid through coordination control of many sets of AC/DC bidirectional flow controllers on the AC & DC hybrid flow sections, achieve real-time and dynamic power balance in the AC & DC hybrid microgrid, and improve frequency quality in the AC microgrid and voltage quality in the DC microgrid. Fig. 6.28 has showed disperse and interactive coordination interface of AC & DC hybrid flow sections.

(3) Comprehensive optimization operation

Distributed resources can be divided into output controllable type and output uncontrollable type according to different control characteristics. Respective applicable control strategies will be adopted in allusion to different types of distributed resources. In allusion to uncertain factors during operation of the AC & DC hybrid microgrid, an optimization algorithm will be used to predict load power, wind power, and PV output power of the microgrid, as well as errors caused by uncertain factors. Output power of batteries will be controllable. Reasonable optimization strategies will be used to optimize the output and ensure safe, stable, reliable, economic, and green operation of the AC & DC hybrid microgrid. Compared with traditional generation means, the economic cost of the new energy microgrid is generally relatively high. However, it provides a huge advantage in terms of energy conservation and environmental protection.

Therefore, certain strategies should be adopted for comprehensive optimization of output of various types of distributed resources to satisfy the objective of economy and environmental protection. A comprehensive optimization operation strategy will take into consideration source load error and zonal

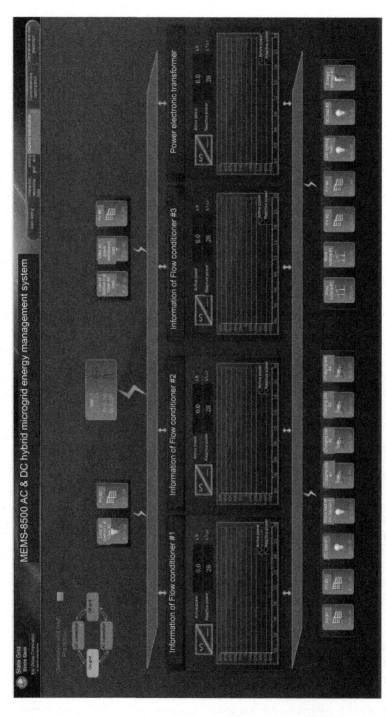

FIG. 6.28 Disperse and interactive coordination between AC & DC hybrid flow sections.

optimization between the AC area and DC area, according to different electrical characteristics of AC area, and DC area in the AC & DC hybrid microgrid, and considering the impact of characteristics of source load on optimization of the AC & DC hybrid microgrid. Moreover, to use complementary advantages of AC and DC power supply means that in the AC & DC hybrid microgrid, interaction between source loads will be used to reduce power loss caused by multiple levels of energy conversion in the microgrid so as to realize the objective of energy conservation and loss reduction. Fig. 6.29 has shown comprehensive optimization operation interface.

(4) Generation prediction and load prediction

As for the AC & DC hybrid microgrid of high capacity penetration, the fluctuation in power generation of distributed resources will impact the grid to some extent, affecting safe and stable operation of the power system directly. Highly efficient power prediction can improve control precision of the PV generation system, which is helpful for the grid dispatching department to plan coordination of PV power generation and general electric sources, arranging operation modes of the grid reasonably and adjusting dispatching plans in time. As for the power system, the impact of the PV power generation system on the grid during grid connection can improve power stability of the system, reduce unfavorable impact caused by connecting PV into the grid effectively, and reduce rotating reserving capacity and operation cost at the same time. Therefore, the AC & DC hybrid microgrid energy management system can conduct short-term and ultra-short-term power prediction of PV power generation, laying a solid foundation for optimizing succeeding algorithms as resources, grid and load of the microgrid, and initial input conditions for comprehensive optimization and operation optimization algorithm.

Load prediction is an important component of the microgrid energy management system. As the demonstration project is applied in enterprise production, the gray correlation method is used in the system to screen original data highly related to prediction data. Then, original load inspection data and corresponding gray correlation data are used for restructuring and are used as a linear regression analysis sample upon standardization. An improved multiple linear regression prediction method based on meteorological and product information is used to predict load of industrial enterprises researched in the demonstration project. Besides this, the impact of key factors on prediction results is taken into consideration. A load prediction method based on meteorological and product information is established based on this to obtain load prediction results in ultra-short-term and short-term scenarios. The prediction results can lay a solid foundation for optimizing succeeding algorithms as resources, grid, and load of the microgrid and initial input conditions for comprehensive optimization and operation optimization algorithms. Fig. 6.30 shows the generation prediction and load prediction interface.

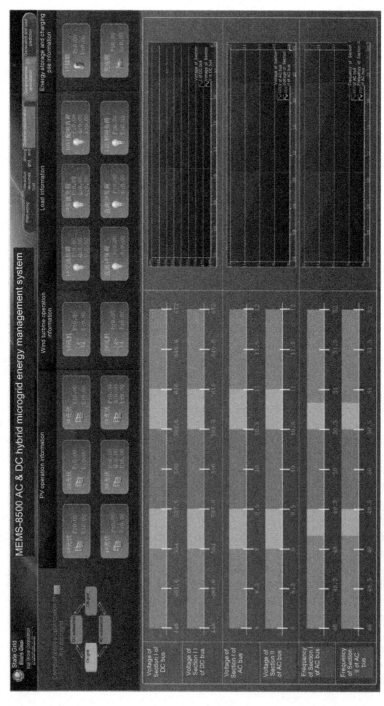

FIG. 6.29 Comprehensive optimization operation.

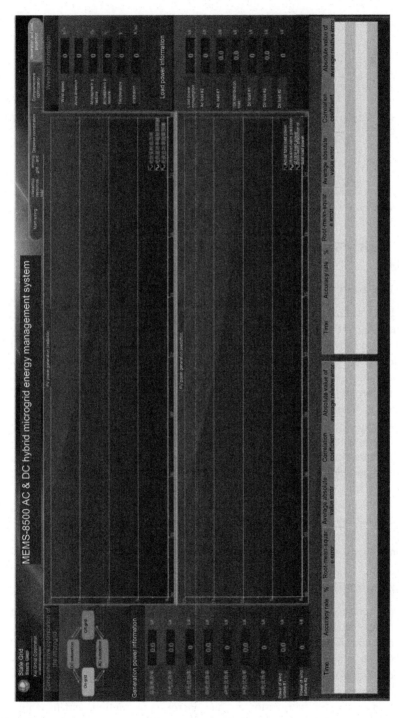

FIG. 6.30 Generation prediction and load prediction.

FIG. 6.31   Aerial view of the Beichen Commercial Center Green Office Demonstration Project.

6.3   Beichen commercial center green office demonstration project

6.3.1   Project overview

Beichen Commercial Center Green Office Demonstration Project is located in the middle of High-end Equipment Industrial Park of Beichen Industry and City Integration Demonstration Area. The project has used roofs of the Commercial Center, newly constructed shed roofs, parking lots, as well as solar energy PV generation, wind power generation, and a microgrid energy storage system constructed around adjacent lake surfaces. Meanwhile, energy management and control as well as a green energy demonstration center are in the building. Newly constructed green energy in the commercial center comprise PV power generation, wind power generation, electric vehicle charging piles, energy storage systems, and so forth. One new energy system is the ground source heat pump system. Fig. 6.31 shows an aerial view of the Beichen Commercial Center Green Office Demonstration Project.

The PV power generation, wind power generation, electric vehicle charging pile, and battery energy storage systems are all electronic power supplies, which have been described in Chapter 2. This section will focus on the introduction of a comprehensive energy management and control platform.

6.3.2   Comprehensive energy management and control platform

(1) The energy management and control platform will plan full life-cycle energy monitoring and management composed of energy production, storage and transportation, application, and reutilization of the Commercial Center as a whole. Sensors are widely distributed in new energy power generation, stored energy charging/discharging, cold and hot load, and other power consumers to establish comprehensive energy monitoring over energy supply, energy storage, and energy consumption, and to realize highly transparent energy consumption.

(2) Wind resources, illumination resources, as well as ground source cold and hot resources will be utilized comprehensively. In addition, organic integration and integrated complementary advantage between electric energy will be given to full play, which can realize the maximum energy utilization of electricity storage, cold storage and heat storage, achieve interaction, complementation and coordinated control among many resources, ensure energy supply and demand balance of buildings, and promote plug-and-play and friendly accessing of clean energy, guaranteeing an energy utilization ratio of 40% for buildings.

(3) Control strategies under indoor temperature self-approximate optimization, energy conservation, equilibrium, user-defined, and other energy consumption modes are established to satisfy personal indoor temperature control under four seasons, working days and nonworking days, working hours and nonworking hours, and other power consumption demands, which guarantees comfortable room temperature and energy cost minimization.

Fig. 6.32 shows a comprehensive first-page interface of energy interaction of the comprehensive energy commanding, management, and control platform.

6.3.3 Function description

6.3.3.1 Energy overview

The energy overview displays overall operation of the entire system, which includes wind, PV, storage, heat pump, power generation, and consumption of buildings, as well as major energy consumption ratio and power supply ratio of buildings. Fig. 6.33 shows an overview interface of energy of the comprehensive energy commanding, management, and control platform.

6.3.3.2 Energy production

Energy production displays the power supply situation and statistical indicators of municipal power, PV, wind power, energy storage, and geothermal energy in the park. Fig. 6.34 shows a comprehensive energy production indicator interface of the comprehensive energy commanding, management, and control platform.

(1) PV power generation displays detailed information on PV power generation in the park, including operation state, generation power, alarm information, meteorological information, overall generation indicators, and so on of each inverter. Fig. 6.35 shows a PV generation interface of energy production of the comprehensive energy commanding, management, and control platform.

(2) Energy storage displays operation information on the energy storage system in the park, including the operation state of PCS, SOC/SOH information, charging/discharging power, alarm information, BMS information, and so

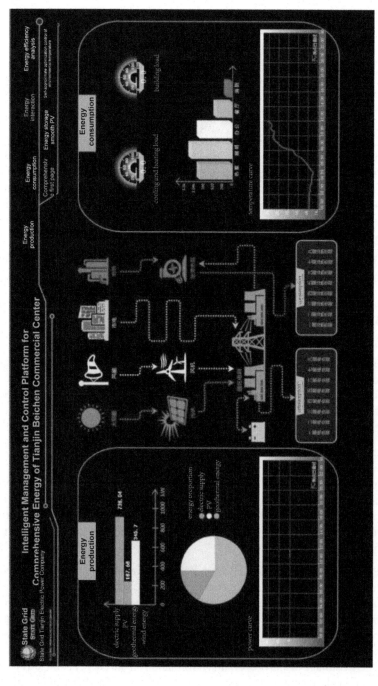

FIG. 6.32 Comprehensive first-page interface of energy interaction of the comprehensive energy commanding, management, and control platform.

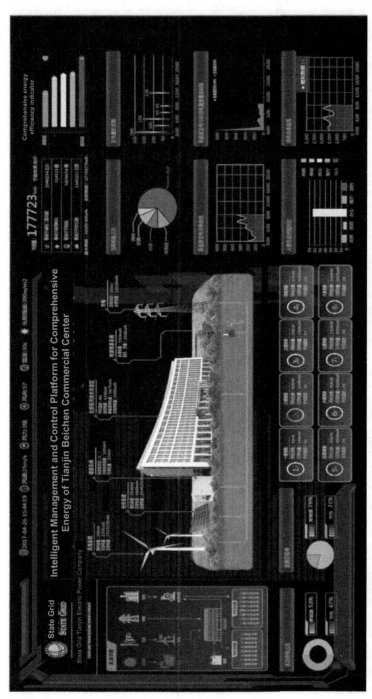

FIG. 6.33 Overview interface for energy of the comprehensive energy commanding, management, and control platform.

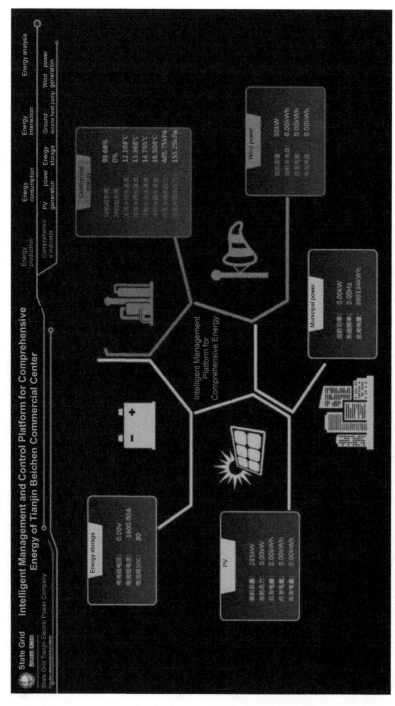

FIG. 6.34 Comprehensive indicator page of energy production of the comprehensive energy commanding, management, and control platform.

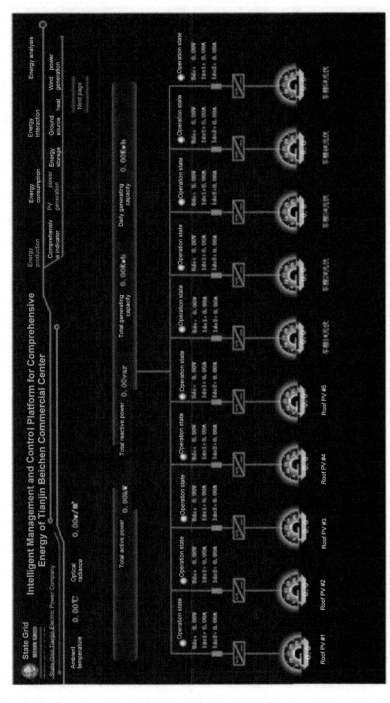

FIG. 6.35 PV power generation interface of energy production of the comprehensive energy commanding, management, and control platform.

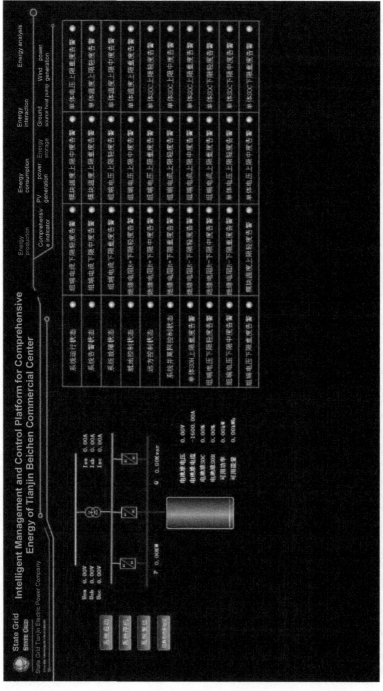

FIG. 6.36 Energy storage interface for energy production of the comprehensive energy commanding, management, and control platform.

on. Fig. 6.36 shows an energy storage interface of energy production of the comprehensive energy commanding, management, and control platform.

(3) The ground source heat pump displays detailed information on the operation of ground source heat pumps in the park, including status information, operation mode, temperature, warning information, and so forth of units. Fig. 6.37 shows a ground source heat pump interface for energy production of the comprehensive energy commanding, management, and control platform.

6.3.3.3   Energy consumption

Energy consumption displays power consumption information in the park, including the consumption ratio of lighting power, power consumption, and ground source heat pumps load (and more), as well as temperature, degree of comfort, and so forth in buildings. Fig. 6.38 shows a comprehensive first-page interface for energy consumption of the comprehensive energy commanding, management, and control platform.

Room energy consumption indicates the power consumption and environment temperature of different areas and floors in the building, where the cooling and heating modes can be controlled in groups. Fig. 6.39 shows the interface of room consumption in the comprehensive energy management and control platform.

6.3.3.4   Smooth control against fluctuation in new energy power generation

This displays the new energy power generation fluctuation by energy storage system in the park. Fig. 6.40 shows the energy storage smooth PV interface for energy interaction of the comprehensive energy commanding, management, and control platform.

6.3.3.5   Environmental temperature self-approximate optimization control

Four different modes are set for environmental temperature in buildings according to applications under different scenes, which are temperature self-approximate optimization control, equilibrium mode, energy conservation mode, and custom mode. As for temperature self-approximate optimization control, comprehensive analysis will be conducted according to peak-valley price and operation condition of ground source heat pumps for the optimized control of temperature in buildings. Fig. 6.41 shows the environmental temperature self-approximate optimization control interface of the comprehensive energy commanding, management, and control platform.

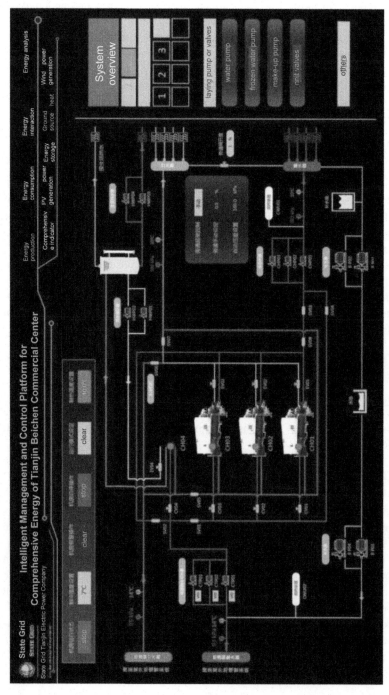

FIG. 6.37 Ground source heat pump interface for energy production of the comprehensive energy commanding, management, and control platform.

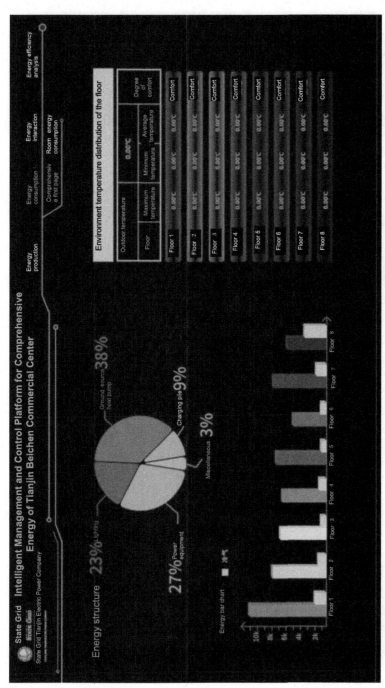

FIG. 6.38   Comprehensive first-page interface of energy consumption of the comprehensive energy commanding, management, and control platform.

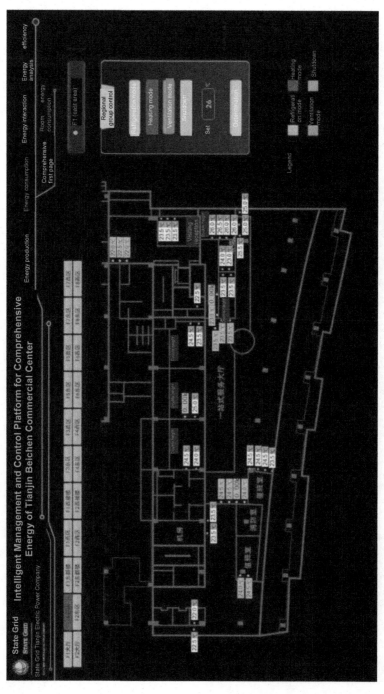

FIG. 6.39 Room energy consumption interface for energy consumption of the comprehensive energy commanding, management, and control platform.

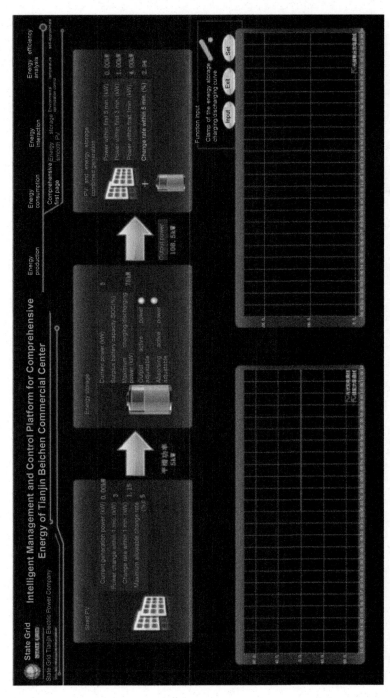

FIG. 6.40 Energy storage smooth PV interface for energy interaction of the comprehensive energy commanding, management, and control platform.

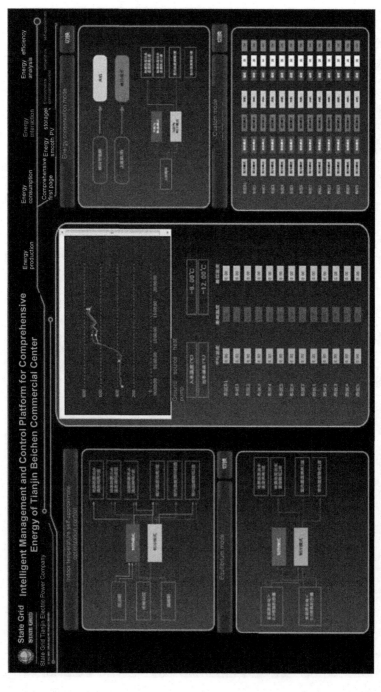

FIG. 6.41 Environmental temperature self-approximate optimization control of the comprehensive energy commanding, management, and control platform.

Annex: Chinese and English contrast of terms

S/N	Abbreviation	Chinese	English full name
1	ADN	主动配电网	Active Distribution Network
2	BMS	电池管理系统	Battery Management System
3	CID	实例配置文件	Configured IED Description
4	CL	可控负荷	Controllable Load
5	CP	渗透率	Capacity Penetration
6	DA	配电自动化	Distribution Automation
7	DER	分布式能源	Distributed Energy Resources
8	DG	分布式发电	Distributed Generation
9	DR	分布式电源	Distributed Resources
10	DS	占优策略	Dominant Strategy
11	DTU	配电终端	Distribution Terminal Unit
12	EMI	电磁干扰	Electromagnetic Interference
13	EPON	以太网无源光网络	Ethernet Passive Optical Network
14	ES	储能	Energy Storage
15	FA	馈线自动化	Feeder Automation
16	FSK	移频键控	Frequency—Shift—Keying
17	FTU	馈线终端	Feeder Terminal Unit
18	GCP	发电量渗透率	Generating Capacity Penetration
19	GIS	地理信息系统	Geographic Information System
20	GOOSE	面向通用对象的变电站事件	Generic Object Oriented Substation Event
21	GT	博弈论	Game Theory
22	HD	电压谐波检测法	Harmonics Detection
23	HMFC	交直流混合微电网潮流控制器	AC & DC Hybrid Microgrid Flow Conditioner
24	IED	智能电子设备	Intelligent Electric Device
25	IS	国际标准	International Standard
26	MG	微电网	Micro-Grid
27	MGCC	微电网控制中心	Micro-Grid Control Center
28	MPPT	最大功率点跟踪	Maximum Power Point Tracking
29	NPC	中点钳位	Neutral Point Clamped
30	OLT	光线路终端	Optical Line Terminal
31	ONU	光网络单元	Optical Network Unit
32	PCC	公共连接点	Point of Common Coupling
33	PCS	能量转换系统	Power Conversion System
34	PDN	被动配电网	Passive Distribution Network
35	PJD	电压相位突变检测法	Phase Jump Detection

S/N	Abbreviation	Chinese	English full name
36	PNP	即插即用	Plug and Play
37	PTP	精密时间协议	Precision Time Protocol
38	PWM	脉冲宽度调制	Pulse Width Modulation
39	SAS	变电站自动化系统	Substation Automation System
40	SHEPWM	滞环控制的特定谐波消除调制	Specific Harmonic Elimination PWM
41	SOC	荷电状态	State of Charge
42	SVPWM	空间矢量脉宽调制	Space Vector Pulse Width Modulation
43	THD	总谐波失真	Total Harmonic Distortion
44	TR	技术报告	Technical Report
45	V2G	电动汽车到电网	Vehicle-to-Grid
46	VFD	电压/频率检测法	Voltage/Frequency Detection
47	VSG	虚拟同步发电机技术	Virtual Synchronous Generator
48	ZVS	零电压开关	Zero Voltage Switch
49	ZVZCS	零电压零电流开关	Zero Voltage Zero Current Switch

Bibliography

[1] R. Li, Development and prospect of plug-and-play of random power supply, Distrib. Util. (1) (2017) 61–67.

[2] R. Li, Concept of plug-and-play of random power supply of cloud one-floor one-end 3-tier architecture, Power Syst. Protect. Control 44 (7) (2016) 47–54.

[3] R. Li, Interconnecting microgrid control technology without communication lines under frequency shift control, Distrib. Util. 33 (10) (2016) 64–70.

[4] X. Li, W. Wang, Research on technical scheme for operation and maintenance of plug-and-play of random resources based on internet of things, Power Syst. Protect. Control 44 (16) (2016) 112–117.

[5] R. Li, B. Guo, Z. Zheng, Detection scheme for injection type active island of low frequency power supplies, Autom. Electr. Power Syst. 44 (5) (2017) 31–36.

[6] R. Li, D. Zhai, Research on automatic voltage regulation method for overvoltage when connecting PV DG into the grid and the microgrid, Power Syst. Protect. Control 43 (22) (2015) 62–68.

[7] D. Zhai, B. Guo, W. Wang, Research on microgrid presynchronization methods based on amplitude and phase position successive approximation, J. Electr. Power (2) (2016) 106–110.

[8] R. Li, X. Li, W. Xie, et al., Technical research on coordination control of the hybrid microgrid based on bus dominance, Distribution Utilization 33 (9) (2016) 73–78.

[9] R. Li, X. Li, B. Wu, et al., Research on function specification of the hybrid microgrid flow conditioners, Smart Grid 4 (9) (2016) 934–940.

[10] R. Li, D. Zhai, B. Guo, et al., Technical research on power conversion, Power Syst. Protect. Control 44 (20) (2016) 24–30.

[11] R. Li, D. Zhai, J. Xu, et al., Research on technical specification for three-level bidirectional DC-DC and AC-DC, Smart Grid 4 (9) (2016) 957–960.

[12] R. Li, Research on differential protection scheme for applicable active grid, Power Syst. Protect. Control 43 (12) (2015) 104–109.

[13] R. Li, A longitudinal protection scheme based on virtual node network topology applicable to active grid of overhead network, Power Syst. Protect. Control 43 (2) (2015) 70–75.

[14] J. Guo, R. Li, X. Li, et al., Research on relay protection of the microgrid and its application, Power Syst. Protect. Control 42 (10) (2014) 135–140.

[15] R. Li, Practices and experiments on key technologies concerning the microgrid, in: Symposium of China's Smart Grid, 2012.

[16] Y. Tian, S. Meng, X. Zou, et al., Research on MW-level island microgrid communication network structure and its engineering application, Power Syst. Protect. Control 43 (19) (2015) 112–117.

[17] Z. Lyu, W. Bingxin, Y. Tian, et al., Research on autonomous operation control methods of MW-level island microgrid and relevant practices, in: Protection and Control Symposium of Relaying Protection Study Committee of Chinese Society for Electrical Engineering, 2015.

[18] R. Li, B. Guo, M. Fu, et al., Research on operation mode switching control of island microgrid and manufacture of devices, in: 2013 National Protection and Control Symposium, 2013.

[19] Y. Li, H. Chen, J. Guo, et al., Research on island detection methods for grid-connected system of distributed generation, Appl. Electronic Tech. 41 (11) (2015) 108–111.

[20] C. Cheng, H. Yang, Z. Zeng, et al., Self-adaptation control methods for rotor inertia of virtual synchronous generators, Autom. Electr. Power Syst. 39 (19) (2015) 82–89.

[21] Z. Lyu, W. Sheng, Q. Zhong, et al., Virtual synchronous generators and their application in the microgrid, Proc. CSEE 34 (16) (2014) 2591–2603.

[22] Y. Ren, IEC 61850 and distribution automation system, Distrib. Util. (05) (2014) 30–33.

[23] W. Feng, H. Lin, Y. Zhang, Design principle for low-voltage anti-island devices in the grid and calculation of parameters, Autom. Electr. Power Syst. 38 (02) (2014) 85–90.

[24] Q. Cheng, Y. Wang, Y. Cheng, et al., Overview research on island detection method in the grid-connected system of distributed generation, Power Syst. Protect. Control 39 (06) (2011) 147–154.

[25] D. Zhang, X. Xu, L. Yang, et al., Impact of distributed resources on overvoltage of the grid, Autom. Electr. Power Syst. 31 (12) (2007) 50–54.

[26] J. Xu, K. Wang, D. Zhai, et al., Flying capacitor type pv power generation grid connection technology based on new-type carrier wave in-phase stratification PWM method, Power Syst. Protect. Control 43 (12) (2015) 134–139.

[27] J. Li, L. Tian, C. Li, Key technologies for distributed grid connection power generation of energy storage combined renewable resources, Electr. Appl. 34 (09) (2015) 28–33.

[28] S. Ye, H. Rui, Z. Liu, et al., Research on bidirectional energy storage converters based on LCL filters, Electr. Power Autom. Equip. 34 (07) (2014) 88–92.

[29] Z. Yao, F. Yu, Q. Zhao, et al., Simulation study on large-scale PV grid-connected system based on modulated multi-level converters, Proc. CSEE 33 (36) (2013) 27–33.

[30] H. Kang, N. Hang, Q. Lu, et al., Analysis on role of distributed pv power generation in smart grid, Power Syst. Clean Energy 29 (10) (2013) 86–90.

[31] B. Zeng, N. Liu, Y. Zhang, et al., Methods for planning double-layer scene of active grid to promote efficient utilization of intermittent distributed resources, Trans. China Electrotech. Soc. 28 (09) (2013) 155–163.

[32] Z. Song, G. Chen, X. He, et al., Research on development of PV power generation and its impact on grid planning, Power Syst. Clean Energy 29 (07) (2013) 92–96.

[33] J. Yang, J. Zhang, X. Gui, Optimization configuration for capacity of hybrid energy system in grid connected PV power generation, Power Syst. Technol. 37 (05) (2013) 1209–1216.

[34] Y. Zhu, J. Yang, L. Gao, et al., Research on reactive optimization planning of grids with asynchronous wind turbines, Power Syst. Protect. Control 40 (05) (2012) 80–84.

[35] L. Chen, B. Zhang, B. Zhou, et al., Manufacture of charging modules of off-board charging generators of electric vehicles, Autom. Electr. Power Syst. 35 (07) (2011) 81–85.

[36] Y. Yi, W. Yao, P. Liu, et al., Manufacture of three-level grid connected converters based on renewable energy, Power Electronic Technol. 44 (06) (2010) 93–94.

[37] Q. Tao, Z. Wu, J. Cheng, et al., Modeling and simulation of the microgrid with PV arrays and fuel cells, Autom. Electr. Power Syst. 34 (01) (2010) 89–93.

[38] C. Liang, X. Duan, Distributed generation and its impact on power system, Autom. Electr. Power Syst. (12) (2001) 53–56.

[39] Y. Zhang, H. Zhang, D. He, et al., Control of the microgrid converters with characteristics of synchronous generators, Trans. China Electrotech. Soc. 29 (07) (2014) 261–268.

[40] X. Zhang, Microgrid operation control solutions and applications, Power Syst. Protect. Control 42 (10) (2014) 141–146.

[41] W. Bao, X. Hu, G. Li, et al., Design of hierarchical control strategy for the microgrid based on synchronous voltage sources, Autom. Electr. Power Syst. 37 (23) (2013) 20–26.

[42] F. Zhou, J. Mao, H. Ma, et al., Discussion on stabilization control of independent island power supply system with distributed resources featuring with high permeability, Power Syst. Protect. Control 41 (02) (2013) 84–90.

[43] M. Ding, X. Yang, J. Su, Control strategy for microgrid inverting power supply based on virtual synchronous generator idea, Autom. Electr. Power Syst. 33 (08) (2009) 89–93.

[44] H. Qin, Wind Turbinegying Wu, S. Peng, et al., Discussion on new operation and maintenance technology for secondary equipment of smart grid, Power Syst. Protect. Control 43 (22) (2015) 35–40.

[45] Z. Li, W. Zhou, X. Qian, et al., Research on power restoration and black start strategy of island in active grid, Trans. China Electrotech. Soc. 30 (21) (2015) 67–75.

[46] X. Huang, J. Weng, Y. Zhang, Evaluation and selection of the microgrid construction planning scheme, Trans. China Electrotech. Soc. 30 (21) (2015) 76–81.

[47] S. Li, R. Wang, Y. Sun, et al., Research on island operation characteristics and island detection test of distributed wind power, Power Syst. Protect. Control 43 (21) (2015) 13–19.

[48] Y. Zhang, F. Yang, J. Zeng, et al., Research on optimized planning scheme for distributed resources in active grid, Power Syst. Protect. Control 43 (15) (2015) 67–72.

[49] G. Xu, Y. Tan, L. Huang, Life evaluation of filled XLPE power cables based on low-rank matrix, Trans. China Electrotech. Soc. 29 (12) (2014) 268–276.

[50] Y. Lǚ, S. Sun, N. Wang, et al., Research and development of on-line warning system for interlocking failures of large-scale wind power bases, Power Syst. Protect. Control 42 (11) (2014) 142–147.

[51] Z. Xiang, Y. Cao, S. Ma, et al., Key points for operation and maintenance of electrical equipment in large-scale PV stations, Sol. Energy 2 (03) (2014) 52–54.

[52] B. Wang, W. Zhao, Research on CIM model based on health state evaluation of transformers, Sci. Technol. Inform. 3 (07) (2014) 248–249.

[53] J. Sun, Z. Wang, Y. Wang, et al., Recovery from failures of complicated grid with distributed resources, Power Syst. Prot. Control 42 (02) (2014) 56–62.

[54] Q. Huo, X. Tang, Technical research on plug-and-play of the microgrid and public supply network, Electr. Power Autom. Equip. 33 (07) (2013) 105–111.

[55] S. Wang, Z. Jiang, J. Zhu, et al., Evaluation of state of grids with distributed resources connected, Power Syst. Protect. Control 41 (13) (2013) 82–87.

[56] X. Huang, J. Zhang, Y. Zhu, et al., Research on monitoring system for electric transmission and transformation equipment based on internet of things, Power Syst. Protect. Control 41 (09) (2013) 137–141.

[57] X. Liu, Discussion on operation and maintenance of large-scale PV grid connection power station, Renew. Energy Resour. 30 (05) (2012) 125–126.

[58] G. Gong, Y. Sun, M. Cai, et al., Research on structure of smart grid oriented internet of things and its application scheme, Power Syst. Protect. Control 39 (20) (2011) 52–58.

[59] X. Li, Q. Gong, H. Qiao, Prospect of application of internet of things in power system, Power Syst. Protect. Control 38 (22) (2010) 232–236.

[60] Y. Luo, L. Shi, G. Tu, et al., Microgrid public information model satisfying requirements for plug-and-play characteristics of distributed resources, Autom. Electr. Power Syst. 34 (08) (2010) 97–100.

[61] B. Wu, Evaluation methods for health state and its application research, Comput. Meas. Control 17 (12) (2009) 2345–2347.

[62] B. Wang, B. Zhang, Z. Hao, Control of emergency load shedding of island microgrid based on power monitoring and frequency change rate, Autom. Electr. Power Syst. 39 (08) (2015) 33–37.

[63] S. Liu, S. Zhuang, M. Xie, Island detection method based on directional deviation control of modulation ratio, Autom. Electr. Power Syst. 39 (03) (2015) 132–139.

[64] Z. Chong, Z. Dai, Y. Jiao, Island detection method of PV grid connection inverters based on injection of zero sequence current, Power Syst. Protect. Control 42 (24) (2014) 65–71.

[65] Z. Chen, X. Meng, X. Wang, et al., New low-frequency phase disturbance method for island detection, Autom. Electr. Power Syst. 38 (13) (2014) 202–207.

[66] J.'g. Liang, X. Jin, X. Wu, et al., Island detection method for microgrid converters based on positive feedback of non-characteristic harmonics, Autom. Electr. Power Syst. 38 (10) (2014) 24–29.

[67] W. Xuezhi, J.'g. Liang, Y. Tong, et al., Methods for evaluating grid impedance based on complex filters and injection of non-characteristic harmonics, Power Syst. Technol. 37 (10) (2013) 2796–2801.

[68] T. Tang, X. Zhang, X. Dong, et al., Research on island detection based on high-frequency injection type impedance detection, Power Electron. Technol. 47 (03) (2013) 70–72.

[69] X. Zhang, R. Wang, X. Liu, et al., Improved active frequency deviation island detection algorithm, Autom. Electr. Power Syst. 36 (14) (2012) 200–204.

[70] J. Kan, S. Xie, Z. Yao, et al., Active frequency deviation island detection technology for grid connection inverters in low-voltage grid, Autom. Electr. Power Syst. 36 (07) (2012) 33–37.

[71] F. Liu, Y. Kang, H. Wang, et al., One improved algorithm for active frequency deviation island detection, Trans. China Electrotech. Soc. 38 (06) (2010) 119–121.

[72] P. Li, L. Zhang, W. Wang, et al., Application and analysis of grid technology, Autom. Electr. Power Syst. 33 (20) (2009) 109–115.

[73] X. Guo, W. Wu, Non-destructive island detection technology without dead zone of the microgrid, Proc. CSEE 29 (25) (2009) 7–12.

[74] Q. Yao, B. Zhao, B. Guo, et al., Injection-type stator ground protection for self-adaptation 20Hz power supply, Autom. Electr. Power Syst. (18) (2008) 71–73.

[75] X. Wang, F. Zhuo, M. Yang, Research on PCC seamless handover control strategy for AC&DC hybrid microgrid, Power Electron. Technol. 46 (08) (2012) 1–3.

[76] L. Zhang, T. Wu, L. Feng, et al., Control for seamless handover of grid-connected system of modulated bidirectional AC/DC converters, Proc. CSEE 32 (06) (2012) 90–96.

[77] C. Zhang, M. Chen, Z. Wang, Research on control strategy for smooth switching of the microgrid operation mode, Power Syst. Protect. Control 39 (20) (2011) 1–5.

[78] Z. Cao, L. Zhou, Y. Zhang, et al., Planning of the microgrid based on power supply stability, Power Syst. Protect. Control 43 (14) (2015) 10–15.

[79] N. Chen, J. Wang, Research, manufacture and application of controllers under microgrid mode, Power Syst. Protect. Control 43 (11) (2015) 115–120.

[80] M. Ding, L. Tian, H. Pan, et al., Research on operation and control strategy for AC&DC hybrid microgrid, Power Syst. Protect. Control 43 (09) (2015) 1–8.

[81] X. Zhang, H. Wang, Shitong Wind turbine, et al., Operation and control of wind power and sea water desalinization isolated microgrid, Power Syst. Protect. Control 43 (04) (2015) 25–31.

[82] Y. Xiong, L. Yu, J. Xu, Connecting PV power generation system into DC microgrid under multiple modes and its control methods, Power Syst. Protect. Control 42 (12) (2014) 37–43.

[83] C. Shen, X. Wu, Z. Wang, et al., Thought on practices and development of the microgrid, Power Syst. Protect. Control 42 (05) (2014) 1–11.

[84] X. Li, B.'e. Li, P. Wang, Analysis on current state and future development of the microgrid technologies, Telecom Power Technol. 32 (05) (2015) 202–207.

[85] G. Li, L. Tan, Z. Wang, et al., Research on impact of single-phase ground failure on circulation current in converters, Power Syst. Protect. Control 44 (03) (2016) 1–7.

[86] Q. Xu, A. Luo, F. Ma, et al., MMC active damping circulation current suppression method considering low-frequency oscillation, Trans. China Electrotech. Soc. 30 (24) (2015) 118–126.

[87] R. Wang, Y. Cheng, S. Sun, et al., Microgrid control based on coordinate rotation virtual impedance and performance analysis, Power Syst. Protect. Control 42 (12) (2014) 78–86.

[88] J. Gao, S. Jianhui, H. Gao, et al., Control strategy for capacitance voltage and circulation current of modulated multiple-level converters, Power Syst. Protect. Control 42 (03) (2014) 56–62.

[89] Y. Yang, Y. Ruan, Y. Tang, et al., Analysis on grid-connected operation circulation current of grid connection inverters in the wind power generation system, High Voltage Technol. 35 (08) (2009) 2012–2018.

[90] W. Tao, J. Li, M. Ding, et al., Development and comparison of grid connection standards of distributed resources, J. Electr. Eng. 11 (04) (2016) 1–8.

[91] J. Zong, K. Bai, H. Liu, et al., Standard on distributed resources and policy overview, North China Electr. Power (09) (2014) 55–60.

[92] W. Bao, X. Hu, G. He, et al., Research on grid connection standards on distributed resources, Power Syst. Technol. 36 (11) (2012) 46–52.

[93] X. Bingyin, Operation, control and protection of 2013 international power supply meeting series report, Autom. Electr. Power Syst. 37 (18) (2013) 1–6.

[94] Mingtian Wind turbine, Z. Zhang, A. Su, et al., Research on feasibility technology for active power distribution system, Proc. CSEE 33 (22) (2013) 12–18.

[95] X. Song, Z. Guo, C. Ni, et al., Research on concentrated network protection of regional grid, Power Syst. Protect. Control 41 (13) (2013) 43–47.

[96] L. Li, X. Zhao, International large grid meeting series report-protection and automation of electric power systems, Autom. Electr. Power Syst. 36 (23) (2012) 1–5.

[97] Mingtian Wind turbine, Overview of the 21st international council on electricity distribution 2011, Distrib. Util. 28 (04) (2011) 8–13.

[98] Mingtian Wind turbine, Research progress and direction for power distribution system and distributed generation sets of 2010 international large grid conference, Power Syst. Technol. 34 (12) (2010) 6–10.

[99] R. Li, Y. He, Zhanfeng Wind turbine, et al., Engineering application practices of three-terminal differential protection for T type transmission lines, Power Syst. Protect. Control 38 (06) (2010) 119–121.

[100] R. Li, A. Yan, Zhanfeng Wind turbine, et al., Engineering application practices of double-loop relay protection with paralleling frames under the same pole, Power Syst. Protect. Control 38 (05) (2010) 82–84.

[101] Mingtian Wind turbine, Overview of the 20th international council on electricity distribution 2009, Distrib. Util. 26 (05) (2009) 16–20.

[102] M. Sun, J. Yu, B. Deng, Analysis on impact of distributed generation on line protection of the microgrid, Power Syst. Technol. 33 (08) (2009) 104–107.

[103] S. Zhao, M. Hu, X. Dou, et al., Research on clock synchronization technology of digital substation based on IEEE1588, Power Syst. Technol. (21) (2008) 97–102.

[104] T. Zhang, A. Yan, Y. Zhao, et al., Research on new relay protection scheme for double loops in the same pole, Relay (08) (2008) 16–19.

[105] Z. Liang, J. Wang, W. Tang, Research on application of EPON technology in smart grid and network design, Telecommun. Electr. Power Syst. 33 (02) (2012) 85–90.

[106] X. Sun, Q. Lü, Coordination control of frequency and voltage of inverters in the low-voltage grid, Trans. China Electrotechn. Soc. 27 (08) (2012) 77–84.

[107] Y. Yao, G. Zhu, X. Liu, Application of the battery energy storage system in improving power quality of the microgrid, Trans. China Electrotechn. Soc. 27 (01) (2012) 85–89.

[108] H. Yang, Y. Gaowang, Zhanfeng Wind turbine, et al., Research and development of the microgrid system controllers and its actual application, Power Syst. Protect. Control 39 (19) (2011) 126–129.

[109] Y. Liu, Z. Wu, Research on calculation method for flow of the microgrid under island operation, Power Syst. Protect. Control 38 (23) (2010) 16–20.

[110] L. Kang, H.G.J. Wu, et al., Discussion on research topics related to distributed resources and when connecting it into the power system, Power Syst. Technol. 34 (11) (2010) 43–47.

[111] L. Su, J. Zhang, L. Wang, et al., Issues related to the microgrid and technical research, Power Syst. Protect. Control 38 (19) (2010) 235–239.

[112] D. Chen, G. Zhu, Characteristics of power transmission in the low-voltage microgrid, Trans. China Electrotech. Soc. 25 (07) (2010) 117–122.

[113] P. Yang, X. Ai, M. Cui, et al., Analysis on economical operation of the microgrid based on particle swarm optimization with multiple types of energy supply system, Power Syst. Technol. 33 (20) (2009) 38–42.

[114] X.'g. Wang, Q. Ai, X. Weihua, et al., Multiobjective optimization of the microgrid energy management with distributed generation, Power Syst. Protect. Control 37 (20) (2009) 79–83.

[115] X. Ai, M. Cui, Z. Lei, Dispatching of environmental protection economy of the microgrid based on chaotic ant swarm algorithm, J. North China Electr. Power Univ. 36 (05) (2009) 1–6.

[116] M. Sun, Y. Zhao, L. Wang, Impact of DG capacity and access mode on relay protection fixed value of the substation, Electr. Power Autom. Equip. 29 (09) (2009) 46–49.

[117] C. Wang, Z. Xiao, S. Wang, Comprehensive control and analysis of the microgrid, Autom. Electr. Power Syst. (07) (2008) 98–103.

[118] W. Huang, J. Lei, X. Xia, et al., Impact of distributed resources on interphase short circuit protection of the microgrid, Autom. Electr. Power Syst. (01) (2008) 93–97.

[119] J. Sheng, L. Kong, Z. Qi, et al., Discussion on research of new grid-microgrid, Relay (12) (2007) 75–81.

[120] M. Ding, M. Wang, Distributed generation technology, Electr. Power Autom. Equip. (07) (2004) 31–36.

[121] Z. Yang, J. Le, K. Liu, et al., Research on grid connection standard of the microgrid, Power Syst. Protect. Control 40 (02) (2012) 66–71.

[122] C. Liu, R. Yuan, B. Liu, et al., Research on operation and development of the microgrid, Southern Power Syst. Technol. 4 (05) (2010) 43–47.

[123] Z. Zhang, R. Yuan, S. Zhao, et al., Discussion on relay protection methods for the microgrid, Power Syst. Protect. Control 38 (18) (2010) 204–209.

[124] Y. Yuan, Z. Li, Y. Feng, et al., Direction and future prospect for developing microgrid in China, Autom. Electr. Power Syst. 34 (01) (2010) 59–63.

[125] W. Zuo, S. Li, X. Wu, et al., Overview of the microgrid technology and its development, Electr. Power 42 (07) (2009) 26–30.

[126] H. Xiao, S. Liu, L. Zheng, et al., Preliminary study on microgrid technology, Power Syst. Protect. Control 37 (08) (2009) 114–119.

[127] Z. Lu, C. Wang, Y. Min, et al., Overview on research of the microgrid, Autom. Electr. Power Syst. (19) (2007) 100–107.

[128] L. Ouyang, X. Ge, Overview on technology of island smart microgrid, Electr. Energy Manage. Technol. (10) (2014) 56–59.

[129] X. Zhang, J. Shu, C. Wu, et al., A type of island distributed PV power generation microgrid, Power Syst. Protect. Control 42 (10) (2014) 56–61.

[130] W. Liu, H. Wang, Y. Zhang, Analysis of characteristics of network messages in the process layer of the intelligent substation and research on communication configuration, Power Syst. Protect. Control 42 (06) (2014) 110–115.

[131] P. Zhang, R. Liu, X. Wang, Research on and design of the microgrid communication system, Comput. Measure. Control 21 (08) (2013) 2209–2212.

[132] P. Xin, P. Yan, Z. Xiao, et al., Research on application of communication network technology in new generation of intelligent substation, Electr. Power Construct. 34 (07) (2013) 17–23.

[133] Q. Zhu, J. Su, Y. Zhao, et al., Analysis of and research on network structure with integration of three networks of intelligent substations, Mech. Electr. Inform. (36) (2012) 142–143.

[134] Wind turbineg Jian, F. Li, W. Xiaoyun, Microcomputer monitoring system based on IEC 61850 standard, Converter Technol. Electr. Traction (02) (2012) 26–29.

[135] J. Zhang, H. Yang, R. Zhao, et al., Overview of research on microgrid communication system, East China Electr. Power 39 (10) (2011) 1619–1625.

Index

Note: Page numbers followed by *f* indicate figures, and *t* indicate tables.

Printed in the United States
By Bookmasters